Shape Memory Alloys 2020

Editor

Gabriel A. López

MDPI • Basel • Beijing • Wuhan • Barcelona • Belgrade • Manchester • Tokyo • Cluj • Tianjin

Editor
Gabriel A. López
University of the Basque
Country UPV/EHU
Spain

Editorial Office
MDPI
St. Alban-Anlage 66
4052 Basel, Switzerland

This is a reprint of articles from the Special Issue published online in the open access journal *Metals* (ISSN 2075-4701) (available at: https://www.mdpi.com/journal/metals/special_issues/shape_memory_alloys_2020).

For citation purposes, cite each article independently as indicated on the article page online and as indicated below:

LastName, A.A.; LastName, B.B.; LastName, C.C. Article Title. *Journal Name* **Year**, *Volume Number*, Page Range.

ISBN 978-3-0365-2470-2 (Hbk)
ISBN 978-3-0365-2471-9 (PDF)

Contents

About the Editor

After his degree in Chemical Engineering at the Nat. Univ. of Comahue (Argentina) Gabriel A. López completed his Ph.D. in Materials Science at the Max Planck Institute for Metals Research in Stuttgart (Germany). Then, having been awarded a Marie Curie Fellowship, he moved to University of the Basque Country UPV/EHU (Bilbao, Spain), where he broadened his expertise in electron microscopy and advanced metallic materials. Since 2016, he has served as Assoc. Professor at UPV/EHU and is head of the Research Group on Thermophysical Properties of Materials.

metals MDPI

Editorial

Shape Memory Alloys 2020

Gabriel A. López

Physics Department, University of the Basque Country UPV/EHU, 48940 Leioa, Spain; gabrielalejandro.lopez@ehu.eus; Tel.: +34-9-4601-8028

Citation: López, G.A. Shape Memory Alloys 2020. *Metals* **2021**, *11*, 1618. https://doi.org/10.3390/met11101618

Received: 28 September 2021
Accepted: 8 October 2021
Published: 12 October 2021

Publisher's Note: MDPI stays neutral with regard to jurisdictional claims in published maps and institutional affiliations.

1. Introduction

Shape memory alloys (SMAs), in comparison to other materials, have the exceptional ability to change their properties, structures, and functionality, depending on the thermal, magnetic, and/or stress fields applied. As is well-known, in recent decades, the development of SMAs has allowed innovative solutions as alternatives in biomedical applications, advanced engineering structures for aerospace and automotive industries, as well as in sensor and actuation systems, among other sectors. Irrespective of this, design and engineering with these special smart materials requires a solid background in materials science in order to consolidate their importance in these fields and to broaden their relevance in other new applications. The goal of this Special Issue is to foster the dissemination of some of the latest research devoted to these special materials from different perspectives.

2. Contributions

Raising the martensitic transformation temperature of SMAs (above 100 °C) is still a challenge, although there are already some materials that are used in different applications. In this Special Issue, Yamabe-Mitarai reviewed TiPd and TiPt-based alloys as important families of high-temperature SMAs [1]. In the context of a detailed investigation it was concluded that multi-component alloys can be good candidates for HT-SMAs, indicating as well that the limitations that need to be overcome entail the suppression of the transformation strain reduction and temperature hysteresis increment. In the search for other alternative high-temperature alloys a great deal of work has been devoted to the study of the CuZr intermetallic. In regards to this system, Biffi et al. [2] introduced interesting work on the effects of Al addition to CuZr-based SMAs in terms of the evolution of the martensitic transformation upon thermal cycling and elucidated important conclusions from a practical point of view.

Other smart alloys that have attracted a great deal of attention in recent years are ferromagnetic SMAs. Among several candidates that have been investigated recently, Co-V-(Si, Al) Heusler alloys are considered an inexpensive SMA for high-temperature applications. In this context, Nakamura et al. [3] provided an interesting investigation of a $Co_{64}V_{15}(Si_{21-x}Al_x)$ alloy and proposed it as a new multifunctional magnetic material. Another remarkable group of ferromagnetic SMAs are the so-called metamagnetic ones and, paying attention to their critical role in material properties, the influence of structural defects in $Ni_{45}Co_5Mn_{37}In_{13}$ alloy was investigated by Pérez-Landazábal et al. [4]. In addition, keeping in mind potential applications, the refrigeration capacity of micro-particles of this alloy as well as the damping properties of the designed SMA-polymer composites have been tested.

It is well-known that the emergence of additive manufacturing technologies has enabled the layer-by-layer production of components. SMAs are not an exception and such techniques have attracted a great deal of interest, although crack formation is still the main challenge. Against this background and paying attention to the latest developments, Ewald et al. [5] innovatively applied laser powder bed fusion to produce crack-free samples of a low-cost Fe-based SMA and achieved a good shape recovery by means of an optimized heat treatment route. Applying a different approach to achieve optimized materials Brailovski et al. [6] demonstrated that simulation-driven processing maps can be used to relate the main laser powder bed fusion parameters to the control of density and grain structure of superelastic Ti-18Zr-14Nb alloys. To complete this section of the Special Issue NiTi, as the most successful SMA, also produced via additive manufacturing could not be absent. Many works have been dedicated to this SMA, and here Biffi et al. [7] provided an experimental comparison of the properties of complex 3D structures and bulk samples, highlighting the main differences.

Characterization techniques are crucial for the development and optimization of new SMAs and production methodologies. In addition to the traditional ones, further advances that shed light on the materials' properties is necessary. In line with this, Sedlak et al. [8] discussed the suitability of laser-based resonant ultrasound spectroscopy (RUS) for the characterization of soft shearing modes in single crystals using three typical examples of SMAs (Cu-Al-Ni, Ni-Mn-Ga, and Ni-Ti), showing special access to high-temperature analysis due to the contactless character of the laser-based arrangement.

To conclude the Special Issue, two very practical situations were introduced as examples of the wide variety of possible SMA applications. On the one hand, taking into account the importance of a precise characterization, Sun et al. [9] clarified the influence of texture type and intensity on the shape memory effect in NiTiNb SMA pipe joints, revealing the causes for the anisotropy of SME via texture changes. Recommendations about the texture effect on the shape memory effect for potential engineering applications were provided. The last contribution, by González et al. [10], investigated the hysteretic behavior and the ultimate energy dissipation capacity of large-diameter NiTi bars subjected to low- and high-cycle fatigue, keeping an eye on the real-life importance of protection from seismic actions. The model was validated with tests conducted on a concrete prototype equipped with large diameter NiTi bars as energy dissipation devices.

Funding: This research was funded by University of the Basque Country, grant number GIU19/019.

Acknowledgments: Support from the University of the Basque Country under the GIU19/019 project is acknowledged.

Conflicts of Interest: The author declares no conflict of interest.

References

1. Yamabe-Mitarai, Y. TiPd- and TiPt-Based High-Temperature Shape Memory Alloys: A Review on Recent Advances. *Metals* **2020**, *10*, 1531. [CrossRef]
2. Biffi, C.A.; Fiocchi, J.; Coduri, M.; Tuissi, A. Effect of Al Addition on Martensitic Transformation Stability and Microstructural and Mechanical Properties of CuZr Based Shape Memory Alloys. *Metals* **2020**, *11*, 1141.
3. Nakamura, K.; Miyake, A.; Xu, X.; Omori, T.; Tokunaga, M.; Kainuma, R. Influence of Structural Defects on the Properties of Metamagnetic Shape Memory Alloys. *Metals* **2021**, *11*, 226.
4. Pérez-Landazábal, J.I.; Sánchez-Alarcos, V.; Ricarte, V.; Lambri, O.H.; Bonifacich, F.G.; Khanna, D.L.R.; Unzueta, I.; García, J.A.; Plazaola, F.; López-García, J.; et al. Influence of Structural Defects on the Properties of Metamagnetic Shape Memory Alloys. *Metals* **2020**, *10*, 1131. [CrossRef]
5. Ewald, F.C.; Brenne, F.; Gustmann, T.; Vollmer, M.; Krooß, P.; Niendorf, T. Laser Powder Bed Fusion Processing of Fe-Mn-Al-Ni Shape Memory Alloy—On the Effect of Elevated Platform Temperatures. *Metals* **2020**, *11*, 185.
6. Brailovski, V.; Kalinicheva, V.; Letenneur, M.; Lukashevich, K.; Sheremetyev, V.; Prokoshkin, S. Control of Density and Grain Structure of a Laser Powder Bed-Fused Superelastic Ti-18Zr-14Nb Alloy: Simulation-Driven Process Mapping. *Metals* **2020**, *10*, 1697. [CrossRef]
7. Biffi, C.C.; Bassani, P.; Fiocchi, J.; Tuissi, A. Microstructural and Mechanical Response of NiTi Lattice 3D Structure Produced by Selective Laser Melting. *Metals* **2020**, *10*, 814. [CrossRef]
8. Sedlák, P.; Janovská, M.; Bodnárová, L.; Heczko, O.; Seiner, H. Softening of Shear Elastic Coefficients in Shape Memory Alloys Near the Martensitic Transition: A Study by Laser-Based Resonant Ultrasound Spectroscopy. *Metals* **2020**, *10*, 1383. [CrossRef]
9. Sun, M.; Fan, Q.; Wang, Y.; Yang, Q.; Chen, J.; Huang, S.; Zhang, Y. {111}<110> Orientation Induced Anisotropy of Shape Memory Effect in NiTiNb Pipe Joints. *Metals* **2020**, *10*, 776. [CrossRef]
10. González-Sanz, G.; Galé-Lamuela, D.; Escolano-Margarit, D.; Benavent-Climent, A. Hysteretic Behavior and Ultimate Energy Dissipation Capacity of Large Diameter Bars Made of Shape Memory Alloys under Seismic Loadings. *Metals* **2019**, *9*, 1099. [CrossRef]

metals

MDPI

Review

TiPd- and TiPt-Based High-Temperature Shape Memory Alloys: A Review on Recent Advances

Yoko Yamabe-Mitarai [1,2]

1 Department of Advanced Materials Science, Graduate School of Frontier Sciences, The University of Tokyo, 5-1-5 Kashiwanoha, Kashiwa-shi, Chiba 277-8561, Japan; mitarai.yoko@edu.k.u-tokyo.ac.jp

2 National Institute for Materials Science, 1-2-1 Sengen, Tsukuba, Ibaraki 305-0047, Japan

Received: 30 September 2020; Accepted: 12 November 2020; Published: 18 November 2020

Abstract: In this paper high-temperature shape memory alloys based on TiPd and TiPt are reviewed. The effect of the alloying elements in ternary TiPd and TiPt alloys on phase transformation and strain recovery is also discussed. Generally, the addition of alloying elements decreases the martensitic transformation temperature and improves the strength of the martensite and austenite phases. Additionally, it also decreases irrecoverable strain, but without perfect recovery due to plastic deformation. With the aim to improve the strength of high-temperature shape memory alloys, multi-component alloys, including medium- and high-entropy alloys, have been investigated and proposed as new structural materials. Notably, it was discovered that the martensitic transformation temperature could be controlled through a combination of the constituent elements and alloys with high austenite finish temperatures above 500 °C. The irrecoverable strain decreased in the multi-component alloys compared with the ternary alloys. The repeated thermal cyclic test was effective toward obtaining perfect strain recoveries in multi-component alloys, which could be good candidates for high-temperature shape memory alloys.

Keywords: high-temperature shape memory alloys; titanium palladium; titanium platinum; multi-component alloys; medium-entropy alloys; high-entropy alloys

1. Introduction

Shape recovery in shape memory alloys (SMAs) occurs during a reverse martensitic transformation from martensite to austenite phases. Thereafter, the SMA operating temperature is related to the martensitic transformation temperature (MTT). High-temperature shape memory alloys (HT-SMAs) are defined as SMAs that can recover their shapes at temperatures above 100 °C. Several applications of HT-SMAs have been proposed. For example, $Ni_{30}Pt_{20}Ti_{50}$, whose MTTs include austenite start temperature, A_s: 262 °C; austenite finish temperature, A_f: 275 °C; martensite start temperature, M_s: 265 °C; and martensite finish temperature, M_f: 240 °C, was applied for active clearance control actuation in the high-pressure turbine section of a turbofan engine [1]. This indicates that the design can offer a small and lightweight package without requiring motion amplifiers that cause efficiency losses and introduce an additional failure mode [1]. Another example is the helical actuators for surge-control applications in helicopter engine compressors [2]. In this application, $Ni_{19.5}Ti_{50.5}Pd_{25}Pt_5$, whose MTTs comprise A_s: 243 °C, A_f: 259 °C, M_s: 247 °C, and M_f: 228 °C, was applied because the alloy exhibited good work capabilities, a 2.5% recoverable strain, and a work output of 9.45 J/cm^3 at 400 MPa [2]. Several SMA applications, such as the active jet engine chevron, springs and wires for a general class of high-temperature actuators, oxygen mask deployment latch, SMA-activated thermal switch for lunar surface applications, variable geometry chevrons, and gas turbine variable area nozzles, have also been proposed [3]. Here, HT-SMAs, $Ni_{19.5}Ti_{50.5}Pd_{25}Pt_5$ or $Ni_{50.3}Ti_{29.7}Hf_{20}$ are used only in springs and wires for a general class of high-temperature actuator. Furthermore, NiTi-based SMAs that can actuate in the temperature range of 70–90 °C are used in other applications.

Raising MTTs is necessary for the development of HT-SMAs. In addition, improving SMA strength is also important because plastic deformation easily occurs at high temperatures, thereby resulting in incomplete shape recovery. Several studies have been conducted to increase MTTs by adding alloying elements such as Hf, Zr, Pd, Pt, and Au, to NiTi [4–10]. Their MTTs successfully increased by adding an alloying element, but a perfect shape recovery was not obtained. Recently, research of NiTi alloys has shifted to $Ni_{50.3}Ti_{29.7}Hf_{20}$, which is strengthened by nano-size precipitates called the "H phase" [11–24]. The austenite finishing temperature A_f of $Ni_{50.3}Ti_{29.7}Hf_{20}$ is typically 166 °C under unloading conditions [13], but it rises to 270 °C under tensile loading conditions at 500 MPa [16]. Furthermore, ageing increased the work output due to the higher transformation strain and the work output under 500 MPa was 16.45 J/cm^3 at A_f of 270 °C [16]. High strength Ni-rich $Ni_{51.2}Ti_{28.8}Hf_{20}$ was also developed and its work output was 23 J/cm^3 under 1700 MPa at A_f of approximately 100 °C, 27 J/cm^3 under 1500 MPa at A_f of approximately 220 °C, and 15 J/cm^3 under 1000 MPa at A_s of 151 °C (A_f was not clearly shown) [19]. The effect of 2000 thermal training cycles under 300 MPa of $Ni_{50.3}Ti_{29.7}Hf_{20}$ was also investigated and it was found that the stable cyclic strain recovery with the almost constant transformation strain [24]. The work output under 300 MPa was approximately 7.5 J/cm^3 at A_f of approximately 220 °C [24].

Another approach to increasing MTTs is using other alloys with MTTs higher than those of NiTi alloys. Therefore, TiPd, TiAu, and TiPt have been studied because they exhibit a martensitic transformation from a B2 to a B19 structure, and their MTT values are higher than 500 °C [25,26]. For example, typical martensitic twin structures were observed in TiPt, whose high potential as an HT-SMA has been established [27,28]. The first investigation on strain recovery at high temperatures was performed for TiPd [29]. A binary TiPd sample was deformed at 500 °C, and the change in its length after it was heated above the A_f was investigated to measure strain recovery [29]. However, only a 10% strain was recovered owing to plastic deformation at 500 °C [29]. The shape recovery behavior of the TiAu alloy was investigated through thermomechanical analysis measurements after compressive deformation at 500 °C [30]. It was found that an 80% strain recovery occurred after a 5% deformation [30]. The effects of both Zr [31] and Ag addition [32] into TiAu on the martensitic transformation and strain recovery was investigated; the MTT decreased by 10% with the addition of Zr and Ag. For example, A_f decreased from 624 °C for TiAu to 511 °C and 594 °C for Ti–50Au–10Zr and Ti–40Au–10Ag, respectively [31,32]. Compressive deformation was applied at a test temperature of 50 °C below M_f, and the deformed samples were heated above A_f to measure the strain recovery. The strain recovery ratios of the Ti–50Au, Ti–50Au–10Zr, and Ti–40Au–10Ag alloys were 68%, 82%, and 76%, respectively [32]. The addition of Zr and Ag was effective in improving the strain recovery of TiAu.

The effects of an alloying element on MTTs, strain recovery, as well as strength of the martensite and austenite phases in TiPd and TiPt alloys, have been investigated by my group [33–50] and are reviewed herein. In addition to the TiPd and TiPt alloys, high- and medium- entropy SMAs (HEAs or MEAs) are also appraised because HEAs and MEAs have been attracting considerable attention as new SMAs. Notably, HEAs and MEAs are multi-component equiatomic or near-equiatomic alloys, which have garnered much interest as new generation structural materials because their high-entropy effects, such as severe lattice distortion and sluggish diffusion, are expected to improve the high-temperature strength of alloys [51,52]. As already shown, it is difficult to achieve perfect strain recovery in HT-SMAs because of the easy introduction of plastic deformation at high temperatures. Furthermore, improvement of strength of SMAs is a key issue for HT-SMAs. Application of HEAs and MEAs to HT-SMAs is expected in this area for which results on multi-component alloys, in particularly MEAs and HEAs, are presented in this paper.

2. Martensitic Transformation Temperature

The MTTs of the TiPd and TiPt alloys measured through differential scanning calorimetry (DSC) (DSC3200, NETZSCH Japan, Yokohama, Japan) are shown in Table 1. The temperature hysteresis,

A_f–M_s, is also summarized in Table 1. Data without reference are from my group's unpublished work. The addition of a third element, such as Hf, Zr, V, Nb, Ta, Cr, Mo, W, Ru, Ir, and Co, was performed for TiPd, while elements, including Ir, Ru, Co, Zr, and Hf, were added for TiPt. In most cases, the MTT decreased with the addition of an alloying element. In Figure 1a, A_f changes with increasing alloying element concentration are shown for some of the ternary alloys. Notably, A_f decreased rather linearly with increasing amounts of alloying elements. The same trend appeared for A_s, M_s, and M_f. The average decrease in M_s per 1 at% of an alloying element up to 5 at% addition is shown in Table 1. Ta and W did not melt well into TiPd, and their particles remained in the alloy, thereby making the matrix composition of $Ti_{45}Pd_{50}W_5$ and $Ti_{45}Pd_{50}Ta_5$ approximately close to that of TiPd. Consequently, the decrease in the MTT by W and Ta was small. For other alloying elements, the decrease in MTT was between 15–37 °C. In Figure 1b, the MTT change with respect to Pd concentration is observed for $Ti_{(95-x)}Pd_xZr_5$; the MTT increased with an increase in the Pd concentration.

Table 1. Martensitic transformation temperature of ternary alloys (°C).

Alloy	Austenite Start Temperature A_s	Austenite Finish Temperature A_f	Martensite Start Temperature M_s	Martensite Finish Temperature M_f	Temperature Hysteresis A_f–M_s	Average Decrease in M_s per 1 at% ΔM_s/at%	Ref.
$Ti_{50}Pd_{50}$	568	587	527	515	60	-	[33]
$Ti_{48}Pd_{50}Hf_2$	538	570	494	472	76	-	
$Ti_{45}Pd_{50}Hf_5$	481	505	460	429	45	−15	[34]
$Ti_{40}Pd_{50}Hf_{10}$	227	306	276	188	30	-	
$Ti_{40}Pd_{55}Hf_5$	528	551	471	487	80	-	[35]
$Ti_{47}Pd_{50}Zr_3$	520	533	491	470	42	-	
$Ti_{45}Pd_{50}Zr_5$	492	508	467	445	41	−15	[33]
$Ti_{43}Pd_{50}Zr_7$	491	506	464	443	42	-	[36]
$Ti_{40}Pd_{50}Zr_{10}$	256	308	221	204	87	-	[33,36]
$Ti_{50}Pd_{45}Zr_5$	324	374	344	312	30	-	
$Ti_{47}Pd_{48}Zr_5$	429	450	370	331	80	-	
$Ti_{43}Pd_{52}Zr_5$	482	529	393	366	136	-	
$Ti_{40}Pd_{55}Zr_5$	530	593	451	438	142	-	[35]
$Ti_{45}Pd_{50}V_5$	508	523	462	440	61	−15	[34]
$Ti_{40}Pd_{55}V_5$	437	559	442	398	117	-	[35]
$Ti_{45}Pd_{50}Nb_5$	486	508	431	408	77	−20	[34]
$Ti_{40}Pd_{55}Nb_5$	487	535	409	360	126	-	[35]
$Ti_{45}Pd_{50}Ta_5$ *	561	579	512	489	67	−4	[34]
$Ti_{45}Pd_{50}Cr_5$	268	356	347	293	9	−37	[34]
$Ti_{45}Pd_{50}Mo_5$	267	383	401	268	−18	−27	[34]
$Ti_{45}Pd_{50}W_5$ *	558	576	513	492	63	−4	[34]
$Ti_{50}Pd_{48}Ru_2$	561	576	516	510	60	-	[37]
$Ti_{50}Pd_{46}Ru_4$	440	482	436	393	46	−23	[37]
$Ti_{50}Pd_{42}Ru_8$	298	354	333	275	22	-	[37]
$Ti_{50}Pd_{37.5}Ru_{12.5}$	-	-	-	-	-	-	[37]
$Ti_{50}Pd_{25}Ru_{25}$	-	-	-	-	-	-	[37]
$Ti_{50}Pd_{48}Ir_2$	556	581	508	500	73	-	[37]
$Ti_{50}Pd_{46}Ir_4$	525	552	510	484	42	−4	[37]
$Ti_{50}Pd_{42}Ir_8$	478	521	479	439	43	-	[37]
$Ti_{50}Pd_{38}Ir_{12}$	471	508	452	395	56	-	
$Ti_{50}Pd_{37.5}Ir_{12.5}$	478	492	452	433	39	-	[37]
$Ti_{50}Pd_{25}Ir_{25}$	-	-	-	-	-	-	[37]
$Ti_{50}Pd_{12.5}Ir_{37.5}$	-	-	-	-	-	-	[37]
$Ti_{50}Pd_{48}Co_2$	523	545	478	464	67	-	[37]
$Ti_{50}Pd_{46}Co_4$	473	493	455	427	38	−18	[37]
$Ti_{50}Pd_{42}Co_8$	375	398	259	327	34	-	[37]
$Ti_{50}Pt_{50}$	1000	1057	989	963	68	-	[38]

Table 1. *Cont.*

Alloy	Austenite Start Temperature A_s	Austenite Finish Temperature A_f	Martensite Start Temperature M_s	Martensite Finish Temperature M_f	Temperature Hysteresis A_f–M_s	Average Decrease in M_s per 1 at% ΔM_s/at%	Ref.
$Ti_{50}Pt_{45}Ir_5$	1050	1081	1030	1008	51	-	[39,40]
$Ti_{50}Pt_{37.5}Ir_{12.5}$	1036	1103	1000	978	103	-	[38]
$Ti_{50}Pt_{25}Ir_{25}$	1121	1190	1110	1068	80	-	[38]
$Ti_{50}Pt_{20}Ir_{30}$	1159	1189	1145	1127	44	-	[39]
$Ti_{50}Pt_{12.5}Ir_{37.5}$	1175	1218	1184	1169	34	-	[38]
$Ti_{55}Pt_{35}Ir_{10}$	931	959	-	-	-	-	[41]
$Ti_{50}Pt_{45}Ru_5$	925	975	913	856	62	−20	[40,42]
$Ti_{45}Pt_{50}Zr_5$	939	985	897	840	88	−23	[40,42]
$Ti_{50}Pt_{45}Co_5$	959	1003	913	855	90	−15	[40,43]
$Ti_{45}Pt_{50}Hf_5$	952	996	905	855	91	−17	[40,44]

*: melting was not perfect.

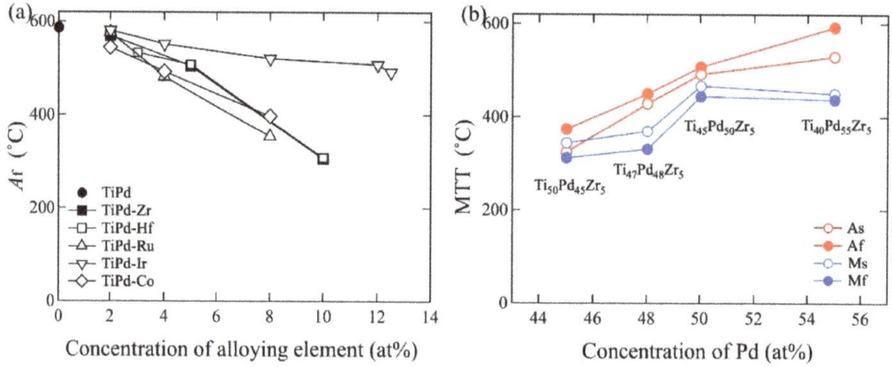

Figure 1. (**a**) Austenite finish temperature change for concentration of alloying element of TiPd [33], TiPd-Zr [33,36], TiPd-Hf [34], TiPd-Ru [37], TiPd-Ir [37], and TiPd-Co [37], and (**b**) martensite transformation temperature for concentration of Pd of $Ti_{50}Pd_{45}Zr_5$, $Ti_{47}Pd_{48}Zr_5$, $Ti_{45}Pd_{50}Zr_5$ [33], and $Ti_{40}Pd_{55}Zr_5$ [35].

The MTT of TiPt was very high; for example, A_s and A_f were 1000 °C and 1057 °C, respectively, as shown in Table 1. Although the addition of Ru, Co, Zr, and Hf decreased the MTT of TiPt, the addition of Ir increased it.

The MTTs of multi-component alloys are summarized in Table 2. Data without reference are from my group's unpublished work. The multi-component alloys were designed based on TiPd as follows: Pd, a group 10 element in the periodic table, was replaced by Co and Ir, which are group 9 elements; by Ni and Pt which are group 10 elements; and by Au, a group 11 element. Furthermore, Ti, a group 4 element, was replaced by Zr and Hf, which are also group 4 elements. Evident DSC peaks were not observed in the three alloys, $Ti_{40}Zr_{10}Pd_{25}Pt_{25}$, $Ti_{45}Zr_5Pd_{37}Ni_{13}$, and $Ti_{45}Zr_5Pd_{25}Pt_{20}Ni_5$, and their MTTs were estimated from the strain–temperature (S–T) curves during the thermal cyclic test under constant stress.

Table 2. Martensitic transformation temperature of multi-element alloys.

Alloy	A_s	A_f	M_s	M_f	A_f–M_s	ΔS_{mix}	Alloy Classification	Ref.
$Ti_{45}Zr_5Pd_{45}Ir_5$	457	475	422	398	53	1.0R	Low-entropy alloy LEA	-
$Ti_{45}Zr_5Pd_{35}Ir_{15}$	313	348	295	256	53	1.2R	Medium-entropy alloy MEA	-
$Ti_{45}Zr_5Pd_{25}Ir_{25}$	-	-	-	-	-	1.2R	MEA	-
$Ti_{45}Zr_5Pd_{45}Pt_5$	497	511	464	439	47	1.0R	LEA	[45]
$Ti_{45}Zr_5Pd_{35}Pt_{15}$	539	557	485	471	72	1.2R	MEA	[45]
$Ti_{45}Zr_5Pd_{25}Pt_{25}$	628	648	571	539	77	1.2R	MEA	[45]
$Ti_{45}Zr_5Pd_{15}Pt_{35}$	725	804	713	590	91	1.2R	MEA	[46]
$Ti_{45}Zr_5Pd_5Pt_{45}$	896	935	774	713	112	1.0R	LEA	[46]
$Ti_{40}Zr_{10}Pd_{25}Pt_{25}$	390	430	245	220	288	1.3R	MEA	[47] ST
$Ti_{45}Zr_4V_1Pd_{50}$	497	508	409	384	99	0.9R	LEA	[36]
$Ti_{45}Zr_{2.5}V_{2.5}Pd_{50}$	523	546	393	368	153	0.9R	LEA	[36]
$Ti_{45}Zr_1V_4Pd_{50}$	496	518	441	424	77	0.9R	LEA	[36]
$Ti_{45}Zr_5Pd_{45}Ni_5$	348	378	328	297	50	1.0R	LEA	[47]
$Ti_{45}Zr_5Pd_{40}Ni_{10}$	216	258	206	177	52	1.1R	MEA	[48]
$Ti_{45}Zr_5Pd_{37}Ni_{13}$	114	230	206	108	24	1.2R	MEA	[47] ST
$Ti_{42}Zr_8Pd_{40}Ni_{10}$	-	-	-	-	-	1.2R	MEA	[47]
$Ti_{45}Zr_5Pd_{30}Ni_{20}$	-	-	-	-	-	1.2R	MEA	[48]
$Ti_{45}Zr_5Pd_{40}Co_{10}$	303	361	339	298	22	1.1R	MEA	[48]
$Ti_{45}Zr_5Pd_{40}Ni_8Co_2$	201	273	212	199	61	1.2R	MEA	[48]
$Ti_{45}Zr_5Pd_{40}Ni_5Co_5$	-	-	-	-	-	1.2R	MEA	[48]
$Ti_{45}Zr_5Pd_{40}Ni_2Co_8$	143	189	135	108	54	1.2R	MEA	[48]
$Ti_{35}Zr_{15}Pd_{20}Pt_{15}Ni_{15}$	-	-	-	-	-	1.5R	High-entropy alloy HEA	[49]
$Ti_{40}Zr_{10}Pd_{20}Pt_{15}Ni_{15}$	-	-	-	-	-	1.5R	HEA	[49]
$Ti_{45}Zr_5Pd_{20}Pt_{25}Ni_5$	559	598	502	432	96	1.3R	MEA	[49]
$Ti_{45}Zr_5Pd_{25}Pt_{20}Ni_5$	390	430	245	220	185	1.3R	MEA	[47] ST
$Ti_{45}Zr_5Pd_{20}Pt_{20}Ni_{10}$	374	442	337	256	105	1.3R	MEA	[49]
$Ti_{35}Zr_{15}Pd_{20}Pt_{15}Au_{15}$	-	-	-	-	-	1.5R	HEA	[50]
$Ti_{35}Zr_{15}Pd_{20}Pt_{15}Co_{15}$	-	-	-	-	-	1.5R	HEA	[50]
$Ti_{45}Zr_5Pd_{25}Pt_{20}Au_5$	520	590	515	453	75	1.3R	MEA	[50]
$Ti_{45}Zr_5Pd_{25}Pt_{20}Co_5$	419	538	431	324	107	1.3R	MEA	[50]
$Ti_{16.7}Zr_{16.7}Hf_{16.7}Ni_{25}Cu_{25}$	184	338	226	112	112	1.6R	HEA	[53]
$Ti_{16.7}Zr_{16.7}Hf_{16.7}Ni_{25}Co_{10}Cu_{15}$	−23	71	36	−80	35	1.8R	HEA	[54]
$Ti_{30}Hf_{20}Pd_{15}Ni_{35}$	537	686	525	479	161	1.3R	MEA	[55]
$Ti_{25}Hf_{25}Pd_{25}Ni_{25}$	680	720	620	580	100	1.4R	MEA	[55]
$Ti_{16.7}Zr_{16.7}Hf_{16.7}Pd_{25}Ni_{25}$	740	780	660	620	120	1.6R	HEA	[55]

Multi-component alloys are classified according to the mixing entropy, ΔS_{mix}, using the following equations [51].

Here, ΔS_{mix} is defined by the following equation:

$$\Delta S_{mix} = -R\sum_{i=1}^{n} x_i ln x_i \quad (1)$$

where x_i is the mole fraction of component i, n is the number of constituent elements, and R is the gas constant (8.314 J/Kmol).

HEA:

$$\Delta S_{mix} \geq 1.5R \quad (2)$$

Medium-entropy alloy (MEA):

$$1.0R \leq \Delta S_{mix} \leq 1.5R \quad (3)$$

Low-entropy alloy (LEA, conventional solid-solution alloy):

$$\Delta S_{\mathrm{mix}} \leq 1.0R \tag{4}$$

The calculated mixing entropies are presented in Table 2. Based on the mixing entropies, the alloy classification is also shown in Table 2. Most quaternary alloys are classified as MEAs, while some are classified as LEAs, i.e., conventional solid-solution alloys. Notably, HEAs have been identified in some multi-component alloys with five or six constituent elements.

The MTTs of the five quaternary alloys, TiZrPdIr, TiZrPdPt, TiZrPdNi, TiZrPdCo, and TiZrVPd, are shown in Table 2. The combination of TiZrPdIr, TiZrPdNi, and TiZrPdCo decreased the MTTs compared with those of $Ti_{45}Zr_5Pd_{50}$. This is because the phase transformations of TiIr, TiNi, and TiCo are different from those of TiPd. For example, TiIr undergoes a two-step phase transformation from the B2 structure at high temperature to a tetragonal structure (distorted B2 structure) at middle temperature, and finally to an orthorhombic structure with space group 65 at low temperature [56]. The MTT depends on Ir concentration. Notably, TiNi undergoes martensitic transformation from the B2 to the B19′ structure [4]. The B2 structure is very stable in TiCo from room temperature to the melting temperature [25]; the addition of these elements is considered to stabilize the B2 structure. However, TiZrPdPt increased the MTT relative to $Ti_{45}Zr_5Pd_{50}$. This is because the same martensitic transformation from B2 to B19 structures, as well as TiPd, occur in TiPt, whose MTT is very high, approximately 1000 °C, as shown in Table 1. Furthermore, Pt and Pd exhibit perfect solubility in both austenite and martensite phases. Consequently, MTT gradually increased with Pt content. In the case of TiZrVPd, the total amounts of Zr and V were maintained at 5 at%, and the values of A_s and A_f were approximately the same as those of $Ti_{45}Zr_5Pd_{50}$, while the M_s and M_f values of TiZrVPd were lower than those of $Ti_{45}Zr_5Pd_{50}$.

Although the MTTs of multi-component alloys depend on the combination of elements, there are some trends whereby those of TiZrPdPtNi and TiZrPdPtAu are relatively high, whereas that of TiZrPdNiCo is relatively low.

The potential of HEAs as SMAs was presented for the first time by Fistov et al., wherein $Ti_{16.7}Zr_{16.7}Hf_{16.7}Ni_{25}Cu_{25}$ was investigated and its A_s and A_f were 184 and 338 °C, respectively [53]. The addition of Co to $Ti_{16.7}Zr_{16.7}Hf_{16.7}Ni_{25}Cu_{25}$ was investigated, but its MTT drastically decreased and A_f was −23 °C [54]. The MTTs of these alloys are summarized in Table 2. The potential of HEAs as HT-SMAs was first shown by Canadinc [55]. The MTT of their alloys was obtained from the DSC curves in reference [55] and is summarized in Table 2. The A_s and A_f of $Ti_{16.7}Zr_{16.7}Hf_{16.7}Pd_{25}Ni_{25}$ were 740 and 780 °C, respectively. In my group's experiments, HEAs and some MEAs did not show martensitic transformation up to 700 °C during DSC measurement. However, martensitic transformation may be possible for these alloys at temperatures higher than 700 °C. The MTT measurement of my group's HEAs is presently undergoing ultra-high temperature DSC. We found that some alloys have a high MTT, approximately close to 1000 °C in multi-component alloys.

Notably, the temperature hysteresis of the multi-component alloys becomes larger relative to that of ternary alloys. The number of alloys with temperature hysteresis exceeding 100 °C and their ratios are summarized in Table 3. The ratio of alloys with temperature hysteresis exceeding 100 °C in ternary alloys is 12% and it increases to 28% in LEAs and to 40% in MEAs and HEAs. It indicates that the ratio of alloys with temperature hysteresis exceeding 100 °C increased with increase of amount of constituent elements in alloys. In multi-component alloys, in particularly in MEAs and HEAs, the amount of constituent elements is equivalent or near-equivalent; therefore, it is expected that severe lattice distortion, which is considered to obstruct martensitic transformation, will occur. In many MEA cases, a drastic decrease in M_s was observed compared with A_f. When SMAs are used as actuators, a smaller temperature hysteresis is necessary to quickly respond to the surrounding environment.

Table 3. Ratio of alloys with temperature hysteresis over 100 °C (%).

Alloys	Number of Alloys with Temperature Hysteresis over 100 °C	Number of Tested Alloys	Ratio of Alloys with Temperature Hysteresis over 100 °C (%)
Ternary alloy in Table 1	5	42	12
LEA in Table 2	2	7	28
MEA and HEA in Table 2	8	20	40

3. Strain Recovery Determined by Compression Test

In the early stage of my group's research [33–35,37–44], the recovery strain was estimated by measuring the sample length before and after the compression test using a strain rate of 3×10^{-3}/s, and after heat treatment above A_f. A sample was deformed by approximately 5% at the test temperature. The deformed sample length (L′) was measured after cooling to room temperature. Thereafter, the sample was re-heated over A_f, cooled to room temperature, and the recovered sample length (L″) was measured. When the initial sample length is L_0, the applied strain ε_a is defined by the following equation:

$$\varepsilon_a = 100 \times ((L_0 - L')/L_0) \tag{5}$$

Recoverable strain ε_r is defined according to the following equation:

$$\varepsilon_r = 100 \times ((L'' - L')/L_0) \tag{6}$$

Strain recovery ratio, i.e., shape memory effect (SME) was evaluated using the following equation:

$$\text{SME (\%)} = \varepsilon_r/\varepsilon_a \times 100 \tag{7}$$

The applied and recoverable strains, as well as the strain recovery ratios of the ternary TiPd and TiPt alloys, are summarized in Table 4. Data without reference are from my group's unpublished work. For the quaternary alloys, this method was used only for TiZrPdIr, and the results are summarized in Table 5. Although it is difficult to compare the effect of alloying elements due to different applied strains, some alloys have a high strain recovery ratio of above 80%. The quaternary alloys also exhibited a high strain recovery ratio of above 80%, as shown in Table 5. In Figure 2, the strain recovery ratios of $Ti_{45}Pd_{50 \times 5}$ are plotted for the periodic table family; it can be observed that the addition of Zr and Hf is effective toward improving strain recovery. Since the atomic size (metallic) of Zr and Hf are 160 and 159 pm, respectively, and larger than those of Ti (147 pm) and Pd (137 pm) [57], the solid-solution hardening effects of Zr and Hf are considered to exceed those of the other alloying elements. Thereafter, the 0.2% flow stress of the austenite and martensite phases were investigated by compression testing using samples with dimensions of $2.5 \times 2.5 \times 5$ mm^3. The 0.2% flow stresses of ternary TiPd and TiPt alloys are summarized in Table 6. Data without reference are from my group's unpublished work. When the martensite phase is deformed, a double yielding behavior often occurs. The first yielding behavior represents the rearrangement of the martensite variant and is referred to as the detwinning stress. The second yielding behavior represents the yield stress of martensite. In Table 6, the detwinning stresses of the martensite phase are also shown. In Figure 3, the 0.2% flow stress of the austenite and martensite phases of $Ti_{45}Pd_{50}X_5$ (X in group 4–6 elements) and $Ti_{50}Pd_{46}Y_4$ (Y in group 8–10 elements) are plotted for the periodic table family. Most alloying elements improved the strength of the austenite and martensite phases compared with those of $Ti_{50}Pd_{50}$. The strengthening behavior of the alloying element on the austenite and martensite phases was similar. Comparing Figures 2 and 3, it is evident that high-strength alloys, such as $Ti_{45}Pd_{50}V_5$, have a lower strain recovery than $Ti_{45}Pd_{50}Zr_5$ and $Ti_{45}Pd_{50}Hf_5$ with a smaller strengthening effect than $Ti_{45}Pd_{50}V_5$. It is difficult to establish the correlation between strength and strain recovery ratio.

Table 4. Strain recovery of ternary alloys.

Alloy	Test Temp., °C	The Difference from Martensite Transformation Temperature, °C	Applied Strain, %	Recoverable Strain, %	Strain Recovery Ratio, %	Ref.
$Ti_{50}Pd_{50}$	538	-	4.1	1.6	40	[33]
$Ti_{50}Pd_{50}$	485	$M_f - 30$	3.8	2.5	67	[34]
$Ti_{50}Pd_{50}$	380	-	3.7	0.46	13	[33]
$Ti_{50}Pd_{50}$	320	-	1.24	0.59	48	
$Ti_{48}Pd_{50}Hf_2$	380	-	2.8	0.4	14	
$Ti_{45}Pd_{50}Hf_5$	456	$A_s - 30$	2.9	2.1	74	
$Ti_{45}Pd_{50}Hf_5$	440	-	2.7	2.3	80	[34]
$Ti_{45}Pd_{50}Hf_5$	380	$M_f - 50$	1.9	1.5	80	
$Ti_{45}Pd_{50}Hf_5$	200	-	5.25	4.1	78	
$Ti_{40}Pd_{50}Hf_{10}$	380	-	3.0	2.4	80	
$Ti_{40}Pd_{55}Hf_5$	457	$M_f - 30$	3.7	1.7	45	[35]
$Ti_{50}Pd_{45}Zr_5$	210	-	4.28	0.7	17	
$Ti_{50}Pd_{45}Zr_5$	295	-	3.1	0.55	18	
$Ti_{50}Pd_{45}Zr_5$	380	-	3.2	1.47	46	
$Ti_{47}Pd_{48}Zr_5$	301	-	4.2	3.6	85	
$Ti_{47}Pd_{48}Zr_5$	399	-	8.3	5.7	69	
$Ti_{47}Pd_{48}Zr_5$	380	-	2.2	1.1	50	
$Ti_{48}Pd_{50}Zr_2$	400	-	6.5	3.4	53	
$Ti_{48}Pd_{50}Zr_2$	380	-	3.1	1.3	43	
$Ti_{47}Pd_{50}Zr_3$	490	$A_s - 30$	3.7	2.6	70	
$Ti_{47}Pd_{50}Zr_3$	415	$M_f - 30$	3.3	2.9	88	
$Ti_{47}Pd_{50}Zr_3$	380	-	3.2	2.8	87	
$Ti_{45}Pd_{50}Zr_5$	415	$M_f - 30$	4.4	3.5	81	[34]
$Ti_{45}Pd_{50}Zr_5$	426	$A_s - 30$	3.1	2.58	84	[33]
$Ti_{45}Pd_{50}Zr_5$	460	-	8	5.2	65	
$Ti_{45}Pd_{50}Zr_5$	380	-	3.5	2,2	62	
$Ti_{45}Pd_{50}Zr_5$	380	-	4.1	3.9	94	[33]
$Ti_{43}Pd_{52}Zr_5$	380	-	2.2	1.75	81	
$Ti_{43}Pd_{52}Zr_5$	451	$A_s - 30$	4.6	4.1	90	
$Ti_{40}Pd_{55}Zr_5$	490	-	7.2	2.0	27	
$Ti_{40}Pd_{55}Zr_5$	450	-	1.2	0.39	34	

Table 4. *Cont.*

Alloy	Test Temp., °C	The Difference from Martensite Transformation Temperature, °C	Applied Strain, %	Recoverable Strain, %	Strain Recovery Ratio, %	Ref.
$Ti_{40}Pd_{55}Zr_5$	380	-	1.9	0.4	22	
$Ti_{43}Pd_{50}Zr_7$	461	$A_s - 30$	1.4	1.2	84	
$Ti_{43}Pd_{50}Zr_7$	413	$M_f - 30$	3.6	3.6	100	
$Ti_{43}Pd_{50}Zr_7$	380	-	3.2	2.9	90	
$Ti_{40}Pd_{50}Zr_{10}$	225	-	7.7	4.8	63	[33]
$Ti_{40}Pd_{50}Zr_{10}$	380	-	1.4	1.4	100	
$Ti_{40}Pd_{55}Zr_5$	408	$M_f - 30$	4	1.2	30	[35]
$Ti_{45}Pd_{50}V_5$	413	$M_f - 30$	2.4	1.2	60	[34]
$Ti_{40}Pd_{55}V_5$	368	$M_f - 30$	3	0.3	10	[35]
$Ti_{45}Pd_{50}Nb_5$	378	$M_f - 30$	3.1	1.1	37	[34]
$Ti_{40}Pd_{55}Nb_5$	330	$M_f - 30$	3	0.3	10	[35]
$Ti_{45}Pd_{50}Ta_5$ *	483	$M_f - 30$	3.0	1.1	35	[34]
$Ti_{45}Pd_{50}Cr_5$	228	$M_f - 30$	3.3	0.4	12	[34]
$Ti_{45}Pd_{50}Mo_5$	256	$M_f - 30$	2.8	0.4	14	[34]
$Ti_{45}Pd_{50}W_5$ *	474	$M_f - 30$	2.3	0.5	22	[34]
$Ti_{50}Pd_{46}Ru_2$	320	-	-	-	40	[37]
$Ti_{50}Pd_{46}Ru_4$	320	-	3.7	2.0	52	
$Ti_{50}Pd_{42}Ru_8$	320	-	3.4	2.5	72	
$Ti_{50}Pd_{48}Ir_2$	320	-	2.6	1.22	47	[37]
$Ti_{50}Pd_{46}Ir_4$	320	-	2.9	1.49	51	[37]
$Ti_{50}Pd_{46}Ir_4$	454	$M_f - 30$	1.0	1.0	100	
$Ti_{50}Pd_{46}Ir_4$	427	$M_f - 60$	2.9	0.55	19	
$Ti_{50}Pd_{42}Ir_8$	320	-	2.17	1.25	57	[37]
$Ti_{50}Pd_{42}Ir_8$	408	$M_f - 30$	1.5	1.3	87	
$Ti_{50}Pd_{42}Ir_8$	427	$M_f - 60$	6.36	2.05	32	
$Ti_{50}Pd_{38}Ir_{12}$	401	$M_f - 30$	1.4	1.01	72	
$Ti_{50}Pd_{48}Co_2$	320	-	2.35	1.36	58	[37]
$Ti_{50}Pd_{46}Co_4$	320	-	3.2	1.4	47	[37]
$Ti_{50}Pd_{42}Co_8$	320	-	3.3	1.0	31	[37]
$Ti_{50}Pt_{50}$	910	$M_f - 50$	6	0	0	[38]
$Ti_{50}Pt_{50}$	850	-	-	-	11	[39]
$Ti_{50}Pt_{45}Ir_5$	850	-	-	-	10	[39,40]
$Ti_{50}Pt_{37.5}Ir_{12.5}$	928	$M_f - 50$	17	2	12	[38]
$Ti_{50}Pt_{37.5}Ir_{12.5}$	850	-	-	-	51	[39]
$Ti_{50}Pt_{25}Ir_{25}$	1018	$M_f - 50$	17	4	23	[38]

Table 4. *Cont.*

Alloy	Test Temp., °C	The Difference from Martensite Transformation Temperature, °C	Applied Strain, %	Recoverable Strain, %	Strain Recovery Ratio, %	Ref.
$Ti_{50}Pt_{25}Ir_{25}$	850	-	-	-	57	[39]
$Ti_{50}Pt_{20}Ir_{30}$	850	-	-	-	36	[39]
$Ti_{50}Pt_{12.5}Ir_{37.5}$	1119	$M_f - 50$	15	0	0	[38]
$Ti_{50}Pt_{12.5}Ir_{37.5}$	850	-	-	-	21	[39]
$Ti_{55}Pt_{35}Ir_{10}$	850	-	4.8	1	20	[41]
$Ti_{56}Pt_{22}Ir_{22}$	850	-	2.5	1.4	56	[41]
$Ti_{58}Pt_{10}Ir_{32}$	850	-	1.9	1.4	73	[41]
$Ti_{50}Pt_{45}Ru_5$	802	$M_f - 50$	2.9	1.3	45	[40,42]
$Ti_{45}Pt_{50}Zr_5$	790	$M_f - 50$	2.1	1.2	58	[40,42]
$Ti_{50}Pt_{45}Co_5$	801	$M_f - 50$	2.5	0.9	36	[40,43]
$Ti_{45}Pt_{50}Hf_5$	806	$M_f - 50$	2.4	1.4	58	[40,44]

*: melting was not perfect.

Table 5. Strain recovery of quaternary alloys.

Alloy	Test Temp., °C	The Difference from Martensite Transformation Temperature, °C	Applied Strain, %	Recoverable Strain, %	Strain Recovery Ratio, %
$Ti_{45}Pd_{45}Ir_5Zr_5$	368	$M_f - 30$	2.8	2.7	96
$Ti_{45}Pd_{35}Ir_{15}Zr_5$	226	$M_f - 30$	2.6	2.2	85

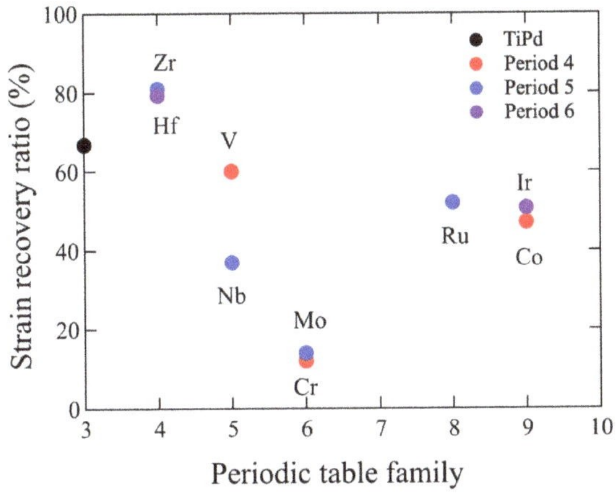

Figure 2. Strain recovery ratio of TiPd [33], $Ti_{45}Pd_{50}X_5$ (X = Zr [33], Hf [34], V [34], Nb [34], Cr [34], and Mo [34]) and $Ti_{50}Pd_{46}Y_4$ (Y = Ru, Co, and Ir) [37] for the periodic table family.

Table 6. Strength of the martensite and austenite phases of the binary and ternary alloys.

Alloy	Test Temp., °C	The Difference from Martensite Transformation Temperature, °C	Detwining Stress, MPa	0.2% Flow Stress, MPa	Ref.
$Ti_{50}Pd_{50}$	617	$A_f + 30$	-	82	[33]
$Ti_{50}Pd_{50}$	538	$A_s - 30$	231	293	[33]
$Ti_{50}Pd_{50}$	485	$M_f - 30$	249	617	
$Ti_{45}Pd_{50}Hf_5$	534	$A_f + 30$	-	267	[34]
$Ti_{45}Pd_{50}Hf_5$	456	$A_s - 30$	152	794	[34]
$Ti_{40}Pd_{55}Hf_5$	581	$A_f + 30$	-	341	[35]
$Ti_{40}Pd_{55}Hf_5$	441	$M_f - 30$	725	1300	[35]
$Ti_{47}Pd_{48}Zr_5$	301	$M_f - 30$	301	1163	
$Ti_{47}Pd_{48}Zr_5$	550	$A_f + 30$	-	628	
$Ti_{47}Pd_{48}Zr_5$	399	$A_s - 30$	302	1449	
$Ti_{48}Pd_{50}Zr_2$	400	$A_s - 30$	631	1504	
$Ti_{48}Pd_{50}Zr_2$	563	$A_f + 30$	-	467	
$Ti_{47}Pd_{50}Zr_3$	563	$A_f + 30$	-	183	
$Ti_{47}Pd_{50}Zr_3$	490	$A_s - 30$	219	666	
$Ti_{47}Pd_{50}Zr_3$	439	$M_f - 30$	234	863	
$Ti_{45}Pd_{50}Zr_5$	538	$A_f + 30$	-	274	[33]
$Ti_{45}Pd_{50}Zr_5$	426	$A_s - 30$	161	722	[33]
$Ti_{45}Pd_{50}Zr_5$	415	$M_f - 30$	298	937	
$Ti_{43}Pd_{50}Zr_7$	536	$A_f + 30$	-	245	
$Ti_{43}Pd_{50}Zr_7$	461	$A_s - 30$	255	733	
$Ti_{43}Pd_{50}Zr_7$	413	$M_f - 30$	227	954	
$Ti_{40}Pd_{50}Zr_{10}$	337	$A_f + 30$	-	386	[33]
$Ti_{40}Pd_{50}Zr_{10}$	225	$A_s - 30$	331	894	[33]
$Ti_{43}Pd_{52}Zr_5$	451	$A_s - 30$	371	1080	
$Ti_{43}Pd_{52}Zr_5$	337	$M_f - 30$	568	1378	
$Ti_{40}Pd_{55}Zr_5$	623	$A_f + 30$	-	386	[35]
$Ti_{40}Pd_{55}Zr_5$	408	$M_f - 30$	1030	1492	[35]
$Ti_{45}Pd_{50}V_5$	591	$A_f + 30$	-	644	[34]
$Ti_{45}Pd_{50}V_5$	413	$M_f - 30$	521	1252	[34]
$Ti_{40}Pd_{55}V_5$	589	$A_f + 30$	643	386	[35]
$Ti_{40}Pd_{55}V_5$	368	$M_f - 30$	744	386	[35]
$Ti_{45}Pd_{50}Nb_5$	625	$A_f + 30$	-	389	[34]
$Ti_{45}Pd_{50}Nb_5$	378	$M_f - 30$	329	836	[34]
$Ti_{40}Pd_{55}Nb_5$	565	$A_f + 30$	601	386	[35]
$Ti_{40}Pd_{55}Nb_5$	330	$M_f - 30$	651	386	[35]
$Ti_{45}Pd_{50}Ta_5$ *	625	$A_f + 30$	-	287	[34]
$Ti_{45}Pd_{50}Ta_5$ *	483	$M_f - 30$	462	941	[34]
$Ti_{45}Pd_{50}Cr_5$	333	$A_f + 30$	-	273	[34]

Table 6. *Cont.*

Alloy	Test Temp., °C	The Difference from Martensite Transformation Temperature, °C	Detwining Stress, MPa	0.2% Flow Stress, MPa	Ref.
$Ti_{45}Pd_{50}Cr_5$	228	$M_f - 30$	317	1000	[34]
$Ti_{45}Pd_{50}Mo_5$	503	$A_f + 30$	-	364	[34]
$Ti_{45}Pd_{50}Mo_5$	256	$M_f - 30$	512	1239	[34]
$Ti_{45}Pd_{50}W_5$ *	615	$A_f + 30$	-	232	[34]
$Ti_{45}Pd_{50}W_5$ *	474	$M_f - 30$	591	1008	[34]
$Ti_{50}Pd_{48}Ru_2$	320	-	-	301	
$Ti_{50}Pd_{46}Ru_4$	520	$A_f + 30$	-	150	
$Ti_{50}Pd_{46}Ru_4$	320	-	196	598	
$Ti_{50}Pd_{46}Ir_4$	580	$A_f + 30$	-	100	
$Ti_{50}Pd_{46}Ir_4$	454	$M_f - 30$	260	740	
$Ti_{50}Pd_{46}Ir_4$	320	-	280	800	
$Ti_{50}Pd_{42}Ir_8$	551	$A_f + 30$	65	150	
$Ti_{50}Pd_{42}Ir_8$	408	$M_f - 30$	320	720	
$Ti_{50}Pd_{38}Ir_{12}$	521	-	215	744	
$Ti_{50}Pd_{38}Ir_{12}$	401	$M_f - 30$	360	850	
$Ti_{50}Pd_{46}Co_4$	520	$A_f + 30$	-	65	
$Ti_{50}Pd_{46}Co_4$	320	-	200	710	
$Ti_{50}Pt_{50}$	1100	$A_f + 50$	-	27	[38]
$Ti_{50}Pt_{50}$	910	$M_f - 50$	-	320	[38]
$Ti_{50}Pt_{37.5}Ir_{12.5}$	1153	$A_f + 50$	-	42	[38]
$Ti_{50}Pt_{37.5}Ir_{12.5}$	928	$M_f - 50$	300	640	[38]
$Ti_{50}Pt_{25}Ir_{25}$	1240	$A_f + 50$	-	35	[38]
$Ti_{50}Pt_{25}Ir_{25}$	1018	$M_f - 50$	230	450	[38]
$Ti_{50}Pt_{12.5}Ir_{37.5}$	1268	$A_f + 50$	-	170	[38]
$Ti_{50}Pt_{12.5}Ir_{37.5}$	1119	$M_f - 50$	285	435	[38]
$Ti_{50}Pt_{45}Ru_5$	-	$A_f + 50$	-	100	[40,42]
$Ti_{50}Pt_{45}Ru_5$	800	$M_f - 50$	355	712	[42]
$Ti_{45}Pt_{50}Zr_5$	-	$A_f + 50$	-	111	[40,42]
$Ti_{45}Pt_{50}Zr_5$	790	$M_f - 50$	600	1468	[40,42]
$Ti_{50}Pt_{45}Co_5$	-	$A_f + 50$	-	72	[40,43]
$Ti_{50}Pt_{45}Co_5$	-	$M_f - 50$	407	523	[40,43]
$Ti_{45}Pt_{50}Hf_5$	-	$A_f + 50$	-	149	[40,44]
$Ti_{45}Pt_{50}Hf_5$	-	$M_f - 50$	565	1113	[40,44]

*: melting was not perfect.

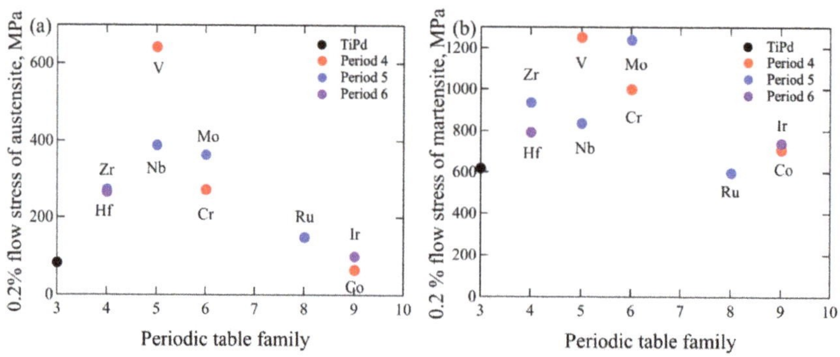

Figure 3. The 0.2% flow stress of the austenite and martensite phases of TiPd [33]: (**a**) $Ti_{45}Pd_{50}X_5$ (X = Zr [33], Hf [34], V [34], Nb [34], Cr [34], and Mo [34]) and (**b**) $Ti_{50}Pd_{46}Y_4$ (Y = Ru, Co, and Ir) [37] for the periodic table family.

4. Strain Recovery Determined by Thermal Cyclic Test under Constant Stress

Recently, strain recovery was investigated through a thermal cyclic compression test under a constant load (Shimadzu AG-X, Shimadzu, Kyoto, Japan). The specimens were heated to 30 °C above A_f and then cooled to a temperature lower than M_f under a constant load of 15–200 MPa. Five thermal cycles under loads of 15, 50, 100, 150, and 200 MPa were applied for the same samples in the order given with heating and cooling rates of 50 °C/min. The work output (work per volume) was obtained from the product of the recovery strain and the applied stress. The strain temperature (S–T) curves of $Ti_{45}Pd_{50}X_5$ (X = Zr, Hf, V, and Nb) are plotted in Figure 4. The transformation strain increased with an increase in the applied stress. The transformation strains of $Ti_{45}Pd_{50}Zr_5$ and $Ti_{45}Pd_{50}Hf_5$ exceed 2% above 50 MPa, as shown in Figure 4a,b. Nonetheless, the transformation strains of $Ti_{45}Pd_{50}V_5$ are less than 1% at an applied stress of 200 MPa. Phase transformation was not evident in TiPdNb, even at 200 MPa. The perfect strain recovery was obtained under a small applied stress of less than 100 MPa in $Ti_{45}Pd_{50}V_5$, but the irrecoverable strain appeared at 150 MPa and increased at 200 MPa. In $Ti_{45}Pd_{50}Zr_5$ and $Ti_{45}Pd_{50}Hf_5$, the small irrecoverable strain was observed, even at 15 MPa, which increased with the applied stress. At 150 and 200 MPa in $Ti_{45}Pd_{50}Hf_5$, trumpet-shaped S–T curves were obtained, as shown by arrows in Figure 4b. This is because the compressive plastic deformation is larger than the expansion of the sample during heating above A_f.

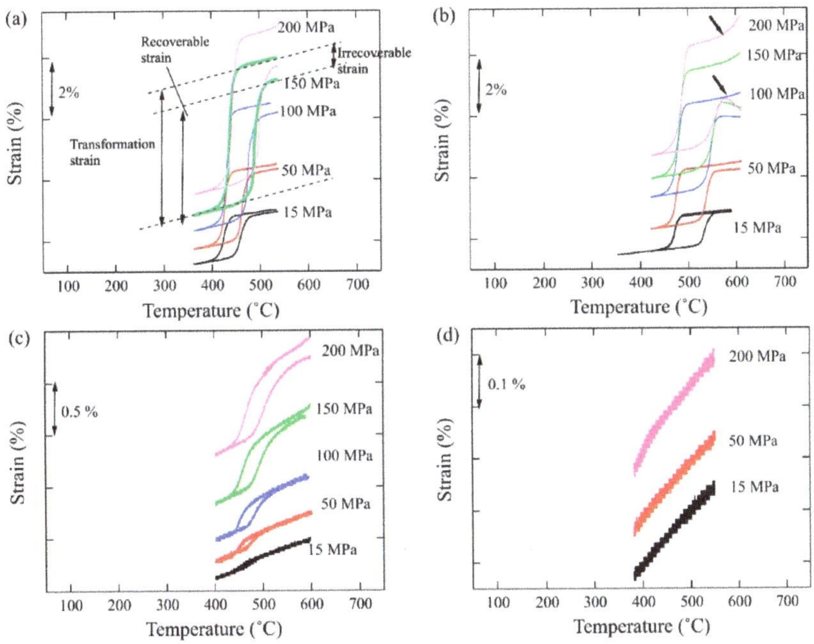

Figure 4. Strain–temperature curves of thermal cycle tests of between 15 and 200 MPa for (**a**) $Ti_{45}Zr_5Pd_{50}$ [45], (**b**) $Ti_{45}Hf_5Pd_{50}$, (**c**) $Ti_{45}V_5Pd_{50}$, and (**d**) $Ti_{45}Nb_5Pd_{50}$.

The recoverable and irrecoverable strains, as well as the work output of $Ti_{45}Pd_{50}X_5$ (X = Zr, Hf, and V) are plotted as a function of the applied stress, as shown in Figure 5; for reference, those of $Ti_{50}Pd_{50}$ at 50 MPa are also plotted [58]. The recoverable strain increased with the applied stress; however, by increasing the irrecoverable strain, the recoverable strain started to decrease after the peak recoverable strain. Although the recoverable strain of $Ti_{50}Pd_{50}$ is similar to that of $Ti_{45}Pd_{50}X_5$ (X = Zr and Hf), the irrecoverable strain is very large compared to that of the ternary alloys. The irrecoverable strain is expected to increase drastically with an increase in the applied stress. It also indicates that the addition of an alloying element is effective toward decreasing the irrecoverable strain. The work output increased with applied stress. Although an irrecoverable strain appeared, a large work output of 7.6 J/cm^3 was obtained for $Ti_{45}Pd_{50}Zr_5$.

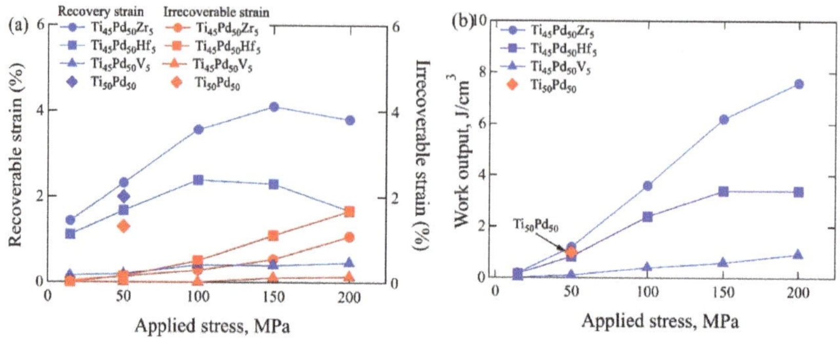

Figure 5. Changes in (**a**) recoverable strain and irrecoverable strain, and (**b**) work output for the applied stress of $Ti_{50}Pd_{50}$ [58], $Ti_{45}Pd_{50}Zr_5$ [45], $Ti_{45}Pd_{50}Hf_5$, and $Ti_{45}Pd_{50}V_5$ derived from the strain–temperature curves shown in Figure 4.

5. Effects of Training

It is known that repeated thermal cyclic tests, which is referred to as "training," reduces the irrecoverable strain, and a perfect recovery is finally obtained [8,59]. For example, the irrecoverable strain of TiNi-based alloys became approximately 0 after 40 cycles under a load of 80 MPa. The training was effective for the TiNi alloys strengthened by the addition of Pd or Sc. Therefore, training was performed for $Ti_{45}Pd_{50}X_5$ (X = Zr and Hf) under a 50-MPa load. The irrecoverable strain vs. the number of thermal cycles is plotted in Figure 6. The irrecoverable strain of $Ti_{45}Pd_{50}Zr_5$ was saturated after 10 cycles, and it remained at approximately 0.1%. The irrecoverable strain of $Ti_{45}Pd_{50}Hf_5$ was approximately double that of $Ti_{45}Pd_{50}Zr_5$, and it was unsaturated. This indicates that the strain recovery of ternary alloys is unstable for repeated cycles. The effect of Zr content on the training effect was investigated for $Ti_{45}Pd_{50}Zr_7$ and $Ti_{45}Pd_{50}Zr_{10}$ [36]. By increasing Zr addition, the irrecoverable strain finally disappeared during training and perfect strain recovery was achieved [36]. This outcome can be attributed to two reasons: (1) the solid-solution hardening effect is larger at high Zr contents, and (2) MTT reduction advantageously affects the suppression of plastic deformation.

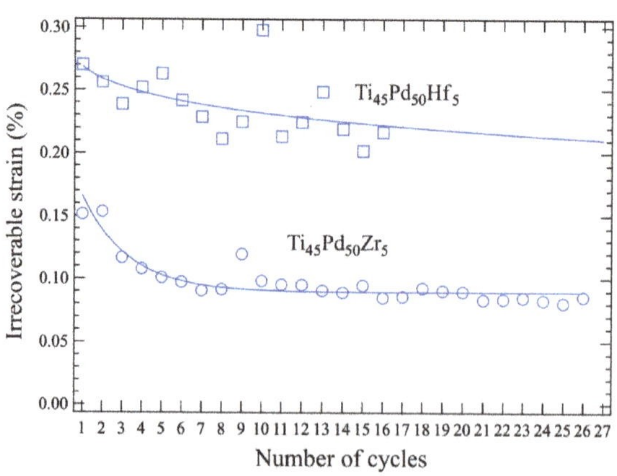

Figure 6. Irrecoverable strain of $Ti_{45}Pd_{50}Zr_5$ and $Ti_{45}Pd_{50}Hf_5$ as a function of number of thermal cycles under 50 MPa.

6. Strain Recovery of Multi-Component Alloys

A thermal cyclic test was performed on the multi-component alloys to investigate strain recovery. Some results have already been published [45–50]. The recoverable and irrecoverable strains, as well as the work output of some of the multi-component alloys are shown in Figure 7; the TiZrPdNi and TiZrPdPt alloys are shown in Figure 7a,c,e, and in Figure 7b,d,f, respectively, while $Ti_{45}Pd_{50}Zr_5$ is shown as a standard sample in all the diagrams. Furthermore, $Ti_{45}Zr_5Pd_{45}Ni_5$, $Ti_{45}Zr_5Pd_{37}Ni_{13}$, $Ti_{40}Zr_{10}Pd_{25}Pt_{25}$, and $Ti_{45}Zr_5Pd_{25}Pt_{20}Ni_5$ are the original data source in this study. In Figure 7a,c,e, the concentrations of Ti and Zr were maintained at 45 and 5 at%, respectively, in TiZrPdNi, and only the concentrations of Pd and Ni were changed. Among the tested alloys in Figure 7a,c,e, TiZrPdNi alloys exhibited a relatively high recoverable strain, although the recoverable strain of $Ti_{45}Zr_5Pd_{40}Ni_{10}$ [48] was similar to that of ternary $Ti_{45}Pd_{50}Zr_5$. Moreover, Ni addition seems to increase the recoverable strain of $Ti_{45}Pd_{50}Zr_5$. The recoverable strain of $Ti_{45}Zr_5Pd_{40}Co_{10}$ [48] was very small, thereby suggesting that Co addition drastically decreased the recovery strain. However, the recoverable strain of the multi-component alloys of TiZrPdNiCo was larger than that of $Ti_{45}Zr_5Pd_{40}Ni_{10}$ [48]. This indicated that the recoverable strain was increased by the effect of the multi-component alloy, i.e., the high-entropy effect. In Figure 7c, the irrecoverable strains of most of the tested alloys were smaller than 0.2%. Only the ternary $Ti_{45}Pd_{50}Zr_5$ and quaternary alloy of $Ti_{45}Zr_5Pd_{45}Ni_5$, which were defined as LEA, represented a large irrecoverable strain exceeding 0.2%. A decrease in the irrecoverable strain of MEAs is also considered to be consequent of the high-entropy effect. As a result of the large recoverable strain, the work output of TiZrPdNi-type multi-component alloys also becomes large, as shown in Figure 7e. The work outputs of some alloys exceeded 10 J/cm^3. The repeated thermal cycling test, i.e., training was also applied for some alloys. For example, training under 300 MPa was performed for $Ti_{45}Zr_5Pd_{40}Ni_{10}$ for 100 cycles [48]. Although the transformation strain decreased, the irrecoverable strain decreased, and a perfect recovery was finally achieved during the thermal cyclic test. The final recovery strain was approximately 3.6% under 300 MPa. Thereafter, the work output was 10.8 J/cm^3 at the A_f of 258 °C. For $Ti_{45}Zr_5Pd_{40}Co_{10}$, training was performed under 700 MPa. After nine cycles, the irrecoverable strain disappeared, and a perfect recovery was achieved. The final recovery strain was approximately 1%, and the work output was 7 J/cm^3 at the A_f of 361 °C.

In Figure 7b,d,f, TiZrPdPt alloys are compared. Again, the concentrations of Ti and Zr were maintained at 45 and 5 at%, respectively, in most of the alloys. Compared with the recoverable strain in TiZrPdNi alloys, as shown in Figure 7a and in TiZrPdPt alloys, as presented in Figure 7b, those of TiZrPdPt alloys were smaller than 3% and less than those of TiZrPdNi alloys. This indicates that the addition of Pt decreases the recoverable strain. The same trend can be observed in quaternary alloys by comparing $Ti_{45}Zr_5Pd_{25}Pt_{25}$ and $Ti_{45}Zr_5Pd_{45}Pt_5$. A small recoverable strain was achieved in the alloy with high Pt content. The effect of Zr on the recoverable strain in quaternary alloys could be understood by comparing $Ti_{40}Zr_{10}Pd_{25}Pt_{25}$ and $Ti_{45}Zr_5Pd_{25}Pt_{25}$, and it was found that Zr addition increased the recoverable strain. In the multi-component alloys, it was found that high Pt addition decreased recoverable strain, as shown in Figure 7b. In TiZrPdPt alloys, it was found that the irrecoverable strain decreased in the multi-component alloys including the MEAs and HEAs, except for the LEA, $Ti_{45}Zr_5Pd_{45}Pt_5$, as shown in Figure 7d. The irrecoverable strain of $Ti_{45}Zr_5Pd_{25}Pt_{25}$ is also small, which may be due to high MTT. The work outputs of the TiZrPdPt alloys are shown in Figure 7f; they are all between 2–6 J/cm^3 and smaller than that of $Ti_{45}Pd_{50}Zr_5$.

Training was performed for $Ti_{45}Zr_5Pd_{20}Pt_{25}Ni_5$ and $Ti_{45}Zr_5Pd_{20}Pt_{20}Ni_{10}$ [49], and perfect recovery was achieved during the thermal cyclic test for approximately 100 cycles under loads of 200, 300, and 400 MPa, whereby the final work outputs were approximately 3.5, 3, and 2 J/cm^3, respectively, for $Ti_{45}Zr_5Pd_{20}Pt_{25}Ni_5$. For $Ti_{45}Zr_5Pd_{20}Pt_{20}Ni_{10}$, final work outputs of 3 and 1.5 J/cm^3 under loads of 200 and 300 MPa were obtained, respectively. The small work output for the large applied stress represents a drastic decrease in the transformation strain. It is necessary to keep the transformation strain of the multi-component alloys during the thermal cyclic test. The thermal cyclic test is considered as a kind of thermal fatigue test and the stable cyclic strain recovery with the constant transformation strain indicates the thermal fatigue life of SMAs and stability as SMA actuators.

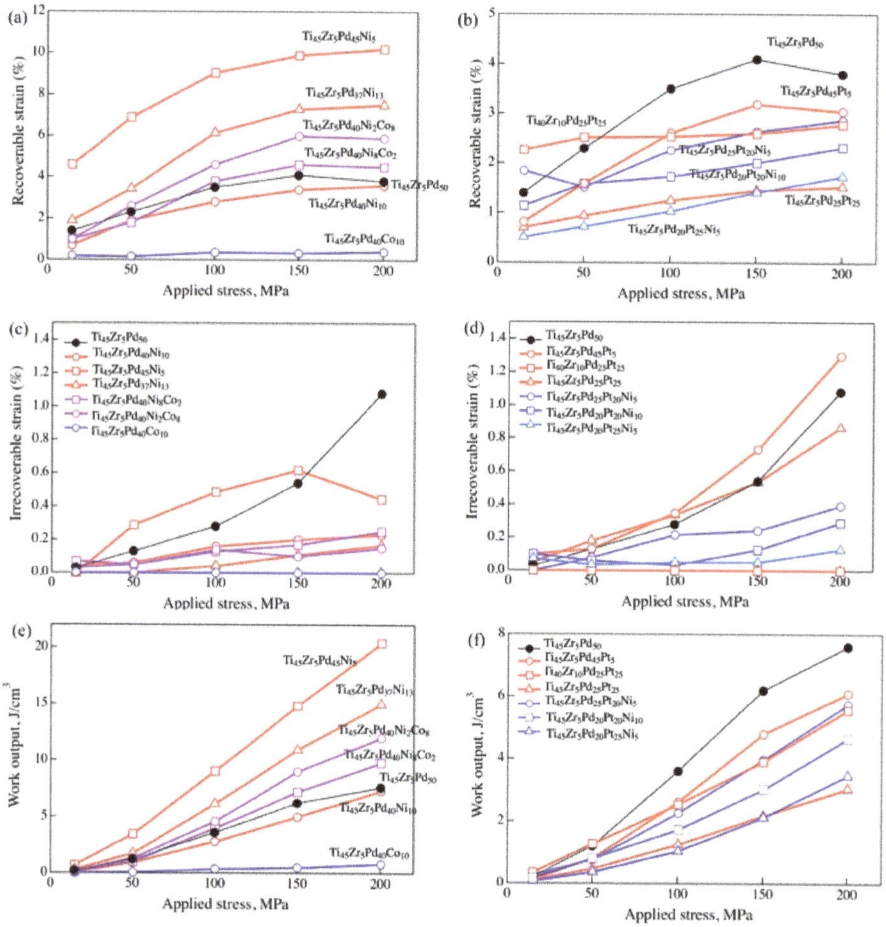

Figure 7. (**a**,**b**) Recoverable strain, (**c**,**d**) irrecoverable strain, and (**e**,**f**) work output obtained from strain–temperature curves of thermal cycle tests of between 15 and 200 MPa for $Ti_{45}Zr_5$ Pd_{50} [45] and multi-component alloys. (**a**,**c**,**d**) TiZrPdNi alloys [47,48], and (**b**,**d**,**f**) TiZrPdPt alloys [45,47,49].

The strength of the austenite and martensite phases of some of the multi-component alloys are investigated, and the results are summarized in Table 7. The strength of the austenite phases in the tested alloys was between 200 and 300 MPa, and they were similar to those of the ternary alloys. This is because the MTTs of the multi-component alloys are relatively high. Therefore, it is difficult to correlate the strength and strain recovery directly. Thereafter, the temperature dependence of the strength of the martensite and austenite phases was investigated for ternary and multi-component alloys [49]. The strength of the multi-component alloys was higher in both martensite and austenite phases when compared at the same temperature. The solid-solution strengthening effect of the multi-component alloys was more evident when the strength of the austenite phase was compared at the same test temperature of 700 °C. The strength of the multi-component alloys was higher than that of the ternary alloys or LEAs.

Table 7. Strength of the martensite and austenite phases of the multi-element alloys.

Alloy	Test Temp., °C	The Difference from Martensite Transformation Temperature, °C	Detwining Stress, MPa	0.2% Flow Stress, MPa	Ref.
$Ti_{45}Pd_{45}Pt_5Zr_5$	-	$A_f + 30$	-	281	[45]
$Ti_{45}Pd_{45}Pt_5Zr_5$	425	$M_f - 30$	246	877	[45]
$Ti_{45}Pd_{35}Pt_{15}Zr_5$	-	$A_f + 30$	-	315	[45]
$Ti_{45}Pd_{35}Pt_{15}Zr_5$	442	$M_f - 30$	387	1080	[45]
$Ti_{45}Pd_{25}Pt_{25}Zr_5$	-	$A_f + 30$	-	231	[45]
$Ti_{45}Pd_{25}Pt_{25}Zr_5$	509	$M_f - 30$	436	1205	[45]
$Ti_{45}Zr_5Pd_{20}Ni_5Pt_{25}$	628	$A_f + 30$	-	322	[49]
$Ti_{45}Zr_5Pd_{20}Ni_5Pt_{25}$	402	$M_f - 30$	-	1266	[49]
$Ti_{45}Zr_5Pd_{20}Ni_{10}Pt_{20}$	472	$A_f + 30$	-	610	[49]
$Ti_{45}Zr_5Pd_{20}Ni_{10}Pt_{20}$	226	$M_f - 30$	-	1615	[49]
$Ti_{45}Zr_5Pd_{25}Pt_{20}Au_5$	620	$A_f + 30$	-	267	[50]
$Ti_{45}Zr_5Pd_{25}Pt_{20}Au_5$	423	$M_f - 30$	-	1169	[50]
$Ti_{45}Zr_5Pd_{25}Pt_{20}Co_5$	568	$A_f + 30$	-	269	[50]
$Ti_{45}Zr_5Pd_{25}Pt_{20}Co_5$	294	$M_f - 30$	-	741	[50]

7. Change in Microstructure before and after Training

The microstructural changes during training were also investigated in $Ti_{43}Zr_7Pd_{50}$ and $Ti_{40}Zr_{10}Pd_{50}$ [36], $Ti_{45}Zr_5Pd_{35}Pt_{15}$ [45], as well as $Ti_{50}Pd_{50}$ [58]. A typical twin structure with multiple variants was observed in the heat-treated sample, but variant selection occurred during the thermal cyclic test, and the major variant was predominantly the [010] axis parallel to the compression axis in all observed alloys. It is suggested that a martensite variant is selected to obtain the largest constriction of the lattice during the compression test. The crystal orientation relationship between the B2 and B19 structures is shown as follows:

$$[110]_{B2}//[100]_{B19},\ [010]_{B2}//[010]_{B19},\ \left[\bar{1}10\right]_{B2}//[001]_{B19}$$

The lattice parameters of the orthorhombic B19 structure of TiPd were a = 0.459, b = 0.28, and c = 0.484 nm [60]. The lattice parameter of b in the B19 structure was the smallest, and the reorientation of the martensite variant along the [010] direction is, therefore, considered to relax the compressive strain under compressive stress.

8. Potential of High-Temperature Shape Memory Alloys (HT-SMAs)

The work outputs obtained by the thermal cyclic test are plotted for A_f, as shown in Figure 8. The open symbols indicate the alloys with imperfect recovery after a single thermal cycle, while solid symbols indicate perfect recovery after training. The green square indicates the work output of $Ni_{50.3}Ti_{29.7}Hf_{20}$ with nano-sized precipitation, which is developed in the USA and is the most expected HT-SMA [16,19]. It is said that the work output of the commercially used TiNi is between 12–18 J/cm^3 for operating temperatures under 100 °C [61]. The work output of $Ni_{50.3}Ti_{29.7}Hf_{20}$ is equivalent to that of TiNi alloys at higher A_f (220 °C) than TiNi. Although the work outputs of some alloys in my group's research exceed 12 J/cm^3 at approximately 200 °C [48], most of them underwent decreasing work outputs with increasing A_f. Nevertheless, it is notable that perfect recovery is achieved at temperatures between 200–600 °C because it is difficult to obtain above 200 °C. The work output obtained at 600 °C was approximately 3 J/cm^3.

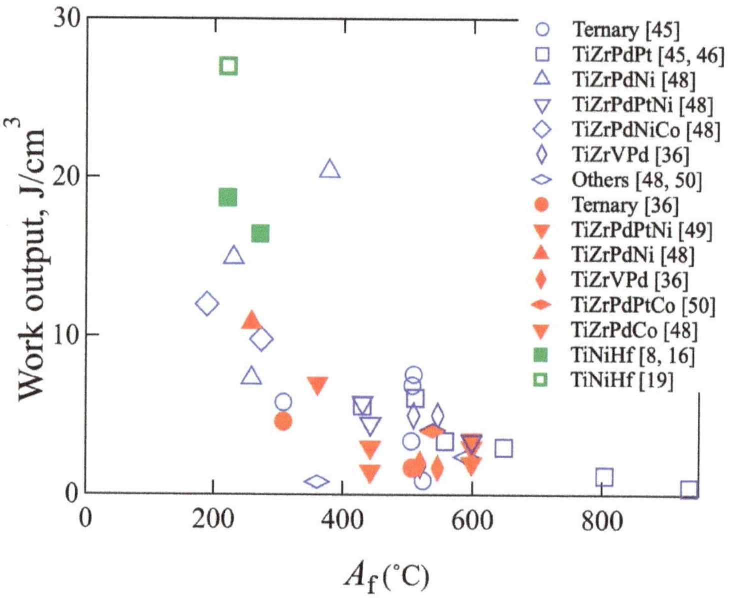

Figure 8. Work output vs. A_f. The open and solid symbols represent the work output of the alloys with imperfect recovery and the work output of the alloys with perfect recovery.

9. Conclusions

This paper reviewed TiPd- and TiPt-based HT-SMAs. The effects of alloying elements on phase transformation, strain recovery, and strength are shown in ternary TiPd and TiPt alloys. In most cases, MTT decreased through the addition of alloying elements, as well as an increase in the alloying element content. The alloying element improved the strength of the martensite and austenite phases by the solid-solution strengthening effect. However, it is difficult to correlate with solid-solution strengthening and strain recovery. A thermal cyclic test was performed to investigate the strain recovery. The irrecoverable strain decreased with the addition of an alloying element, but it was difficult to achieve perfect recovery, even after repeated thermal cyclic tests (training). Multi-component alloys were also investigated. Notably, MTT can be controlled by a combination of the constituent elements. For example, Zr, Ni, and Co decreased MTT, whereas Pt increased MTT. The effect of alloying elements on the recoverable strain differed from that of the MTT. For example, Ni increased the recoverable strain, while Pt and Co decreased it. The extra solid-solution hardening effect relative to ternary alloys was found in the multi-component alloys. Consequently, the irrecoverable strains were smaller in the multi-component alloys than those in the ternary alloys. In many cases, training was effective in achieving perfect recovery in the multi-component alloys; however, a decrease in the transformation strain also occurred. Multi-component alloys can be good candidates for HT-SMAs; the limitations that need to be overcome entail the suppression of the transformation strain reduction and temperature hysteresis increment.

Funding: The study was partly supported by "Precious Metals Research Grant of TANAKA Memorial Foundation", for which the author expresses her thanks.

Acknowledgments: This work was performed in cooperation with M. Kawakita, A. Wadood, and R. Arockiakumar as a part of their Post-doctoral research; W. Tasaki and W. Takebe as a part of their Bachelors research; H. Sato and H. Matsuda as a part of their Bachelors and Masters research; and B. Ohl, K. Bogdanowicz, E. Muszalska, D. Kuczyska, A. Wierzchowska, A. Chmielewska, K. Aimizu, and M. Yamamoto as a part of their internship program.

Conflicts of Interest: The author declares no conflict of interest.

References

1. DeCastro, J.A.; Melcher, K.J.; Noebe, R.D. NASA/TM-2005-213834, AIAA-2005-3988. Available online: https://ntrs.nasa.gov/citations/20050203852 (accessed on 30 September 2020).
2. Stebner, A.; Padula, S., II; Noebe, R.; Lerch, B.; Quinn, D. Development, Characterization, and Design Considerations of $Ni_{19.5}Ti_{50.5}Pd_{25}Pt_5$ High-temperature Shape Memory Alloy Helical Actuators. *J. Intell. Mater. Syst. Struct.* **2009**, *20*, 2107–2126. [CrossRef]
3. Benefan, O.; Brown, J.; Calkins, F.T.; Kumar, P.; Stebner, A.P.; Turner, T.L.; Vaidyanathan, R.; Webster, J.; Young, M.L. Shape memory alloy actuator design: CASMART collaborative best practices and case studies. *Int. J. Mech. Mater. Des.* **2014**, *10*, 1–42. [CrossRef]
4. Ma, J.; Karaman, I.; Noebe, R.D. High temperature shape memory alloys. *Int. Mater. Rev.* **2010**, *55*, 257–315. [CrossRef]
5. Bigelow, G.S.; Padula, S.A., II; Garg, A.; Gaydosh, D.; Noebe, R.D. Characterization of ternary NiTiPd high-temperature shape-memory alloys under load-biased thermal cycling. *Metall. Mater. Trans. A* **2010**, *41*, 3065–3079. [CrossRef]
6. "Development of a Numerical Model for High-Temperature Shape Memory Alloys" NASA Technical Reports Server, 2013. Available online: https://ntrs.nasa.gov/citations/20060028493 (accessed on 30 September 2020).
7. Atli, K.C.; Karaman, I.; Noebe, R.D.; Garg, A.; Chumlyakov, Y.I.; Kireeva, I.V. Improvement in the Shape Memory Response of $Ti_{50.5}Ni_{24.5}Pd_{25}$ High-Temperature Shape Memory Alloy with Scandium Microalloying. *Metall. Mater. Trans. A* **2010**, *41*, 2485–2497. [CrossRef]
8. Atli, K.C.; Karaman, I.; Noebe, R.D.; Maier, H.J. Comparative analysis of the effects of severe plastic deformation and thermomechanical training on the functional stability of $Ti_{50.5}Ni_{24.5}Pd_{25}$ high-temperature shape memory alloy. *Scr. Mater.* **2011**, *64*, 315–318. [CrossRef]
9. Atli, K.C.; Karaman, I.; Noebe, R.D. Work output of the two-way shape memory effect in $Ti_{50.5}Ni_{24.5}Pd_{25}$ high-temperature shape memory alloy. *Scr. Mater.* **2011**, *65*, 903–906. [CrossRef]
10. Atli, K.C.; Karaman, I.; Noebe, R.D.; Garg, A.; Chumlyakov, Y.I.; Kireeva, I.V. Shape memory characteristics of $Ti_{50.5}Ni_{24.5}Pd_{25}Sc_{0.5}$ high-temperature shape memory alloy after severe plastic deformation. *Acta Mater.* **2011**, *59*, 4747–4760. [CrossRef]
11. Yang, F.; Coughlin, D.R.; Phillips, P.J.; Yang, L.; Devaraj, A.; Kovarik, L. Structure analysis of a precipitate phase in an Ni-rich high-temperature NiTiHf shape memory alloy. *Acta Mater.* **2013**, *61*, 3335–3346. [CrossRef]
12. Santamarta, R.; Arroyave, R.R.; Pons, J.; Evrigen, A.; Karaman, I.; Karacka, H.E.; Noebe, R.D. TEM study of structural and microstructural characteristics of a precipitate phase in NI-rich Ni-Ti-Hf and Ni-Ti-Zr shape memory alloys. *Acta Mater.* **2013**, *61*, 6191–6206. [CrossRef]
13. Bigelow, G.S.; Garg, A.; Padula, S.A., II; Gaydosh, D.J.; Noebe, R.D. Load-biased shape-memory and superelastic properties of a precipitation strengthened high-temperature $Ni_{50.3}Ti_{29.7}Hf_{20}$ alloy. *Scr. Mater.* **2011**, *64*, 725–728. [CrossRef]
14. Coughlin, D.R.; Phillips, P.J.; Bigelow, G.S.; Garg, A.; Noebe, R.D.; Mills, M.J. Characterization of the microstructure and mechanical properties of a 50.3Ni-29.7Ti-20Hf shape memory alloy. *Scr. Mater.* **2012**, *67*, 112–115. [CrossRef]
15. Evirgen, A.; Basner, F.; Karaman, I.; Noeve, R.D.; Pons, J.; Santamarta, R. Effect of Aging on the martensitic transformation characteristics of a Ni-rich NiTiHf high temperature shape meory alloys. *Funct. Mater. Lett.* **2012**, *5*, 1250038. [CrossRef]
16. Karaca, H.E.; Saghaian, S.M.; Ded, G.; Tobe, H.; Basaran, B.; Maier, H.J.; Noebe, R.D.; Chumlyakov, Y.I. Effects of nanoprecipitation on the shape memory and material properties of an Ni-rich NiTiHf high temperature shape memory alloy. *Acta Mater.* **2013**, *61*, 7422–7431. [CrossRef]
17. Benafan, O.; Garg, A.; Noebe, R.D.; Bigelow, G.S.; Padula, S.A.; Gaydosh, D.J.; Schell, N.; Mabe, J.H.; Vaidyanathan, R. Mechanical and functional behavior of Ni-rich $Ni_{50.3}Ti_{29.7}Hf_{20}$ high temperature shape memory alloy. *Intermet* **2014**, *50*, 94–107. [CrossRef]
18. Evirgen, A.; Karaman, I.; Santamarta, R.; Pons, J.; Hayrettin, C.; Noebe, R.D. Relationship between crystallographic compatibility and thermal hysteresis in Ni-rich NiTiHf and NiTiZr high temperature shape memory alloys. *Acta Mater.* **2016**, *121*, 374–383. [CrossRef]

19. Saghaian, S.M.; Karaca, H.E.; Tobe, H.; Turabi, A.S.; Saedi, S.; Saghaian, S.E.; Chumlyakov, Y.I.; Noebe, R.D. High strength NiTiHf shape memory alloys with tailorable properties. *Acta Mater.* **2017**, *134*, 211–220. [CrossRef]
20. Amin-Ahmadi, B.; Gallmeyer, T.; Pauza, J.G.; Duerig, T.W.; Noebe, R.D.; Stebner, A.P. Effect of a pre-aging treatment on the mechanical behaviors of $Ni_{50.3}Ti_{49.7-x}Hf_x$(x < 9 at.%) Shape memory alloys. *Scr. Mater.* **2018**, *147*, 11–15.
21. Amin-Ahmadi, B.; Pauza, J.G.; Shamimi, A.; Duerig, T.W.; Noebe, R.D.; Stebner, A.P. Coherency strains of H-phase precipitates and their influence on functional properties of nickel-titanium-hafnium shape memory alloys. *Scr. Mater.* **2018**, *147*, 83–87. [CrossRef]
22. Evirgen, A.; Pons, J.; Karaman, I.; Santamarta, R.; Noebe, R.D. H-Phase Precipitation and Martensitic Transformation in Ni-rich Ni-Ti-Hf and Ni-Ti-Zr High-Temperature Shape Memory Alloys. *Shap. Mem. Superelasticity* **2018**, *4*, 85–92. [CrossRef]
23. Amin-Ahmadi, B.; Noebe, R.D.; Stebner, A.P. Crack propagation mechanisms of an aged nickel-titanium-hafnium shape memory alloy. *Scr. Mater.* **2019**, *159*, 85–88. [CrossRef]
24. Karakoc, O.; Atli, K.C.; Evirgen, A.; Pons, J.; Santamarta, R.; Benafan, O.; Noebe, R.D.; Karaman, I. Effects of training on the thermomechanical behavior of NiTiHf and NiTiZr high temperature shape memory alloys. *Mater. Sci. Eng. A* **2020**, *794*, 139857. [CrossRef]
25. ASM Alloy Phase Diagrams Center; Villars, P.; Okamoto, H.; Cenzual, K. (Eds.) ASM International, Materials Park, OH, 2006. Available online: http://www1.asminternational.org/AsmEnterprise/APD (accessed on 4 September 2020).
26. Donkersloot, H.C.; van Vocht, J.H.N. Martensitic transformations in gold-titanium, palladium-titanium and platinum-titanium alloys near the equiatomic composition. *J. Less Common Met.* **1970**, *20*, 83–91. [CrossRef]
27. Biggs, T.; Witcomb, M.J.; Cornish, L.A. Martensite-type transformations in platinum alloys. *Mater. Sci. Eng. A* **1999**, *273–278*, 204–207. [CrossRef]
28. Biggs, T.; Corite, M.B.; Witcomb, M.J.; Cornish, L.A. Martensitic Transformations, Microstructure, and Mechanical Workability of TiPt. *Metall. Trans. A* **2003**, *32A*, 2267–2272. [CrossRef]
29. Otsuka, K.; Oda, K.; Ueno, Y.; Piao, M.; Ueki, T.; Horikawa, H. The shape memory effect in a $Ti_{50}Pd_{50}$ alloy. *Scr. Met. Mater.* **1993**, *29*, 1355–1358. [CrossRef]
30. Declairieux, C.; Denquin, A.; Ochin, P.; Poriter, R.; Vermaut, P. On the potential of $Ti_{50}Au_{50}$compound as a high temperature shape memory alloy. *Intermetallics* **2011**, *19*, 1461–1465. [CrossRef]
31. Wadood, A.; Hosoda, H.; Yamabe-Mitarai, Y. Phase transformation, osidation and shape memory properties of Ti-50Au-10Zr alloy for high temperature applications. *J. Alloys Compd.* **2014**, *595*, 200–205. [CrossRef]
32. Wadood, A.; Yamabe-Mitarai, Y. Silver- and Zirconium-added ternary and quaternary TiAu based high temperature shape memory alloys. *J. Alloys Compd.* **2015**, *646*, 1172–1177. [CrossRef]
33. Kawakita, M.; Takahashi, M.; Takahashi, S.; Yamabe-Mitarai, Y. Effect of Zr on phase transformation and high-temperature shape memory effect in TiPd alloys. *Mater. Lett.* **2012**, *89*, 336–338. [CrossRef]
34. Yamabe-Mitarai, Y. Development of High-Temperature Shape Memory Alloys above 673 K. *Mater. Sci. Forum* **2017**, *879*, 107–112. [CrossRef]
35. Arockiakumar, R.; Takahashi, M.; Takahashi, S.; Yamabe-Mitarai, Y. Microstructure, mechanical and shape memory properties of Ti-55Pd-5x (x = Zr, Hf, V, Nb) alloys. *Mater. Sci. Eng. A* **2013**, *585*, 86–93. [CrossRef]
36. Sato, H.; Kim, H.Y.; Shimojo, M.; Yamabe-Mitarai, Y. Training effect on microstructure and shape recovery in Ti-Pd-Zr alloys. *Mater. Trans.* **2017**, *10*, 1479–1486. [CrossRef]
37. Yamabe-Mitarai, Y.; Wadood, A.; Arockiakumar, R.; Hara, T.; Takahashi, M.; Takahashi, S.; Hosoda, H. High-Temperature Shape Memory Alloys Based on Ti-Platinum Group Metals compounds. *Mater. Sci. Forum* **2014**, *783–786*, 2541–2545. [CrossRef]
38. Yamabe-Mitarai, Y.; Hara, T.; Miura, S.; Hosoda, H. Mechanical properties of Ti-50(Pt, Ir) high-temperature shape memory alloys. *Mater. Trans.* **2006**, *47*, 650–657. [CrossRef]
39. Yamabe-Mitarai, Y.; Hara, T.; Miura, S.; Hosoda, H. Phase transformation and shape memory effect of Ti(Pt, Ir). *Metall. Trans. A* **2012**, *43A*, 2901–2911. [CrossRef]
40. Yamabe-Mitarai, Y.; Arockiakumar, R.; Wadood, A.; Suresh, K.S.; Kitashima, T.; Hara, T.; Shimojo, M.; Tasaki, W.; Takahashi, M.; Takahashi, S.; et al. Ti(Pt, Pd, Au) based high temperature shape memory alloys. *Mater. Today* **2015**, *2S*, S517–S522. [CrossRef]

41. Yamabe-Mitarai, Y.; Hara, T.; Kitashima, T.; Miura, S.; Hosoda, H. Composition dependence of phase transformation behavior and shape memory effect of Ti(Pt, Ir). *J. Alloys Compd.* **2013**, *577S*, S399–S403. [CrossRef]
42. Wadood, A.; Takahashi, M.; Takahashi, S.; Hosoda, H.; Yamabe-Mitarai, Y. High-temperature mechanical and shape memory properties of TiPt-Zr and TiPt-Ru alloys. *Mater. Sci. Eng. A* **2013**, *564*, 34–41. [CrossRef]
43. Wadood, A.; Yamabe-Mitarai, Y. TiPt-Co and TiPt-Ru high temperature shape memory alloys. *Mater. Sci. Eng. A* **2014**, *610*, 106–110. [CrossRef]
44. Wadood, A.; Yamabe-Mitarai, Y. Recent research and developments related to near-equiatomic titanium-platinum alloys for high-temperature applications. *Platinum Metals Rev.* **2014**, *2*, 61–67. [CrossRef]
45. Tasaki, W.; Shimojo, M.; Yamabe-Mitarai, Y. Thermal cyclic properties of Ti-Pd-Pt-Zr high-temperature shape memory alloys. *Crystals* **2019**, *9*, 595. [CrossRef]
46. Yamabe-Mitarai, Y.; Takebe, W.; Shimojo, M. Phase transformation and shape memory effect of Ti-Pd-Pt-Zr high-temperature shape memory alloys. *Shap. Mem. Superelasticity* **2017**, *3*, 381–391. [CrossRef]
47. Yamabe-Mitarai, Y.; Tasaki, W.; Ohl, B.; Sato, H. Effect of Ni, Co, and Pt on Phase transformation and shape recovery of TiPd high temperature shape memory alloys. In Proceedings of the International Conference on Processing & Manufacturing of Advanced Materials, Paris, France, 9–13 July 2018.
48. Yamabe-Mitarai, Y.; Ohl, B.; Bogdanowicz, K.; Muszalska, E. Effect of Ni and Co on Phase Transformation and Shape Memory Effect of Ti-Pd-Zr alloys. *Shap. Mem. Superelasticity* **2020**, *6*, 170–180. [CrossRef]
49. Matsuda, H.; Sato, H.; Shimojo, M.; Yamabe-Mitarai, Y. Improvement of high-temperature shap-memory effect by multi-component alloyin for TiPd alloys. *Mater. Trans.* **2019**, *11*, 2282–2291. [CrossRef]
50. Matsuda, H.; Shimojo, M.; Murakami, H.; Yamabe-Mitarai, Y. Martensitic Transformation of High-Entropy and Medium-Entropy Shape Memory Alloys. *Mater. Sci. Forum* **2021**, *1016*, 1802–1810.
51. Yeh, J.W.; Chen, S.K.; Lin, S.J.; Gan, J.Y.; Chin, T.S.; Shun, T.T.; Tsau, C.H.; Chang, S.Y. Nanostructured high-entropy alloys with multiple principal elements: Novel alloy design concepts and outcomes. *Adv. Eng. Mater.* **2004**, *6*, 299–303. [CrossRef]
52. Canter, B.; Chang, I.T.H.; Knight, P.; Vincent, A.J.B. Microstructural development in equiatomic multicomponent alloys. *Mater. Sci. Eng. A* **2004**, *375–377*, 213–218. [CrossRef]
53. Firstov, G.S.; Kosorukova, T.A.; Koval, Y.N.; Odnosum, V.V. High entropy shape memory alloys. *Mater. Today* **2015**, *2S*, S499–S504. [CrossRef]
54. Chen, C.-H.; Chen, Y.-J. Shape memory characteristics of $(TiZrHf)_{50}Ni_{25}Co_{10}Cu_{15}$ high entropy shape memory alloy. *Scr. Mater.* **2019**, *162*, 185–189. [CrossRef]
55. Canadinc, D.; Trehern, W.; Ma, J.; Karaman, I.; Sun, F.; Chaudhry, Z. Ultra-high temperature multi-component shape memory alloys. *Scr. Mater.* **2019**, *158*, 83–87. [CrossRef]
56. Yamabe-Mitarai, Y.; Hara, T.; Phasha, M.; Ngoepe, P.; Chikwanda, H. Phase transformation and crystal structure of IrTi. *Intermetallics* **2012**, *31*, 26–33. [CrossRef]
57. Available online: https://en.wikipedia.org/wiki/Atomic_radii_of_the_elements_(data_page) (accessed on 1 September 2020).
58. Hisada, S.; Matsuda, M.; Yamabe-Mitarai, Y. Shape Change and Crystal Orientation of B19 Martensite in Equiatomic TiPd Alloy by Isobaric Test. *Metals* **2020**, *10*, 375. [CrossRef]
59. Atli, K.C.; Franco, E.B.; Karaman, I.; Gaydosh, D.; Noebe, R.D. Influence of crystallographic compatibility on residual strain of TiNi based shape memory alloys during thermo-mechanical cycling. *Mater. Sci. Eng. A* **2013**, *574*, 9–16. [CrossRef]
60. *Pearson's Crystal Data—Crystal Structure Database for Inorganic Compounds (on CD-ROM)*; Villars, P.; Cenzual, K. (Eds.) ASM International: Materials Park, OH, USA, 2010.
61. Karaca, H.E.; Acar, E.; Tobe, H.; Saghaian, S.M. NiTiHf-based shape memory alloys. *Mater. Sci. Technol.* **2014**, *30*, 1530–1544. [CrossRef]

metals MDPI

Article

Effect of Al Addition on Martensitic Transformation Stability and Microstructural and Mechanical Properties of CuZr Based Shape Memory Alloys

Carlo Alberto Biffi [1,*], Jacopo Fiocchi [1], Mauro Coduri [2] and Ausonio Tuissi [1]

1 National Research Council, Institute of Condensed Matter Chemistry and Technologies for Energy, Via Previati 1/E, 23900 Lecco, Italy; jacopo.fiocchi@icmate.cnr.it (J.F.); ausonio.tuissi@cnr.it (A.T.)
2 Department of Chemistry, University of Pavia, Viale Taramelli 16, 27100 Pavia, Italy; codurimauro@gmail.com
* Correspondence: carloalberto.biffi@cnr.it

Abstract: In this work, the effect of the Al content (x = 5, 10, and 15 at. %) on the martensitic transformation (MT) and microstructure and mechanical properties of $Cu_{(50-x)}Zr_{50}Al_x$ alloys was studied. The microstructure of the alloys was characterized at room temperature by means of scanning electron microscopy and X-ray diffraction. An increase in Al content reduces the amount of transforming CuZr phase, and consequently the secondary phase formation is favored. The evolution of the MT upon thermal cycling was investigated as a function of the Al content by differential scanning calorimetry. MT temperatures and enthalpies were found to be decreased when increasing the Al content. Al addition can induce a sudden, stable MT below 0 °C, while the binary alloy requires ten complete thermal cycles to stabilize. Finally, the mechanical properties were investigated through microhardness and compression testing. No linear dependence was found with composition. Hardness lowering effect was observed for 5–10 at. % of Al content, while the hardness was increased only for 15 at. % Al addition with respect to the binary alloy. Similarly, compressive response of the alloys showed behavior dependent on the Al content. Up to 10 at. % Al addition, the alloys indicate a superelastic response at room temperature, while higher Al content induced untimely failure.

Keywords: shape memory alloys; intermetallic; microstructure; differential scanning calorimetry; X-ray diffraction; mechanical testing

Citation: Biffi, C.A.; Fiocchi, J.; Coduri, M.; Tuissi, A. Effect of Al Addition on Martensitic Transformation Stability and Microstructural and Mechanical Properties of CuZr Based Shape Memory Alloys. *Metals* **2021**, *11*, 1141. https://doi.org/10.3390/met11071141

Academic Editors: Sergey Kustov and Alexander V. Shelyakov

Received: 17 May 2021
Accepted: 1 July 2021
Published: 20 July 2021

1. Introduction

In recent years, worldwide interest has spurred on great efforts in the development of shape memory alloys (SMAs) to be applied for automotive, aerospace, mechanical, and control systems [1]. Among the SMAs, the near-equiatomic NiTi compound is the most applied for industrial applications, as it shows stable and performing functional properties [1,2]. Other SMA systems, even Ni and Ti free ones, have been investigated in the literature because of the possibility of offering performances with limited costs within different operating temperature ranges. Among these, the intermetallic CuZr has been studied extensively, since it exhibits a reversible, non-thermoelastic martensitic transformation (MT) above 100 °C when subjected to thermal cycles. The literature offers some studies reporting investigations into the microstructure and functional properties of both CuZr [3–9] and CuZr based shape memory alloys [10–18].

At first, Firstov et al. reported that equiatomic, intermetallic CuZr shows a MT below 140 °C [3] from the parent austenite (A) phase (cubic B2, space group Pm-3m), stable at high temperature, into two martensitic phases (M) with monoclinic structure (space groups Cm and *P*21/*m*) [4,5]. This system can show secondary phases, Cu10Zr7 and CuZr2, as reported by Carvalho et al. [6].

Biffi et al. showed that the characteristic temperatures of the MT at the first thermal cycle and the Cu/Zr ratio in quasi-equiatomic composition are strongly correlated [7]. It was reported that the equiatomic composition appears to be the best compromise between the greatest amount of material involved in the MT and the highest transformation temperatures for considering CuZr within high temperature shape memory alloys (HTSMAs) In other studies, it was found that upon thermal cycling the initial MT, present at the first thermal cycle, is suppressed, and a new one, stable at temperatures below 0 °C and with limited thermal hysteresis, takes its place [8–10]. Electrical resistivity measurements showed a large shift of the characteristic temperatures, followed, after tens of thermal cycles, by a new type of MT at approximately −10 °C [8]. Moreover, thermal cycling through differential scanning calorimetry evidenced this new MT, from austenite to a new martensitic structure, which was different than the original one present in the first few cycles [9,10].

In yet another study, it was found that a B2 CuZr solid-phase transformation can exist at a high temperature [11]. The CuZr-based alloys were classified into three different types, related to the precipitation and stability of the B2 phase. An innovative way based on a prediction strategy for the formation of different sized CuZr-based shape memory bulk metallic glass composites (BMGC) was proposed for evaluating the competition among the vitrification, the precipitation of metastable B2 CuZr, and room-temperature equilibrium phases.

Due to this lack of stability during phase transformations, as well as to improve the alloys' workability, efforts have been made to study the addition of several alloying elements, like Co, Cr, Ni, Al, and Ti [10–16].

It was found that Co addition has a stabilizing effect on the MT temperatures during thermal cycling [10–14], while Ni addition can shift the MT to higher temperatures above 200 °C but weakens the shape memory effect [10–13,15].

Hot workability can also be improved by the addition of Ni and Co together, while Ti alloying does not confer any evident benefits to Cu-Zr systems and reduces the transformation temperatures [16,17]. Other alloying elements, like Cr and Al, generally decrease the characteristic temperatures of the MT [14,18–23]. In particular, the effect of Al addition up to 15 at. % in place of Cu was studied [18–22]: it was found that the transformation temperatures of the ternary alloys were decreased with respect to the binary one at the first thermal cycle [20].

In the present work, the effect of Al addition to CuZr based SMAs is investigated in terms of the evolution of the MT upon thermal cycling. A comparison with the equiatomic alloy, thermally cycled for stabilizing the MT [9], was made in order to define the principal differences among the binary and ternary systems. The study was carried out through differential scanning calorimetry (DSC) for measuring the characteristic temperature of the MT. Scanning electron microscopy (SEM) with energy dispersion spectroscopy (EDS) and θ–2θ X-ray Diffraction analysis (XRD), all at room temperature, were proposed for the microstructural analysis. Mechanical properties were evaluated using microhardness measurements and compression testing, which were also done at room temperature.

2. Experimental

Pure metals Zr, Al, and OFHC grade Cu were melted by means of a vacuum arc furnace (Leybold mod. LK 6/45) with a non-consumable tungsten electrode to obtain small cylindrical buttons (30 mm in diameter and about 40 g in weight). The melting furnace was equipped with a water-cooled copper crucible to avoid contamination of the liquid pool; the melting was done in a high purity inert atmosphere (Ar 99.99% at 400 mbar pressure). The nominal atomic compositions of the chosen alloy were the following: $Cu_{(50-x)}Zr_{50}Al_x$ with x = 0, 5, 10, 15 at. %. The produced buttons were re-melted six times, being flipped over after each melting step in order to ensure high homogeneity in their composition. After melting, the specimens were fully annealed at 850 °C for 3 h under Ar atmosphere (800 mbar) with final quenching in water at room temperature for homogenizing the microstructure of the small ingots. The water quench allows us to perform uniform cooling in the entire sample. The grain size was evaluated to be around 100–200 μm.

The microstructure of the alloys was observed by means of scanning electron microscopy (SEM Leo 1413, Carl Zeiss AG, Jena, Germany) using back scattered electrons (BSE); quantitative analysis of the secondary phases was performed by means of energy dispersive X-ray spectroscopy (EDS), calibrated with a Co reference. Diffraction data (CuKα, λ = 1.5418 Å) were collected on a θ:2θ vertical scan Panalytical X'Pert PRO diffractometer (Malvern Panalytical Ltd, UK), equipped with parallel (Soller) slits—0.04 rad—and a real time multiple strip detector. The generator was operated at 40 kV and 30 mA; slits were used with a divergence of 0.5°. The scans were performed in the 25–55° 2θ range at 25 °C on bulk specimens. The goal of the XRD analysis is to identify the phases present. Since the specimens are textured, we used Rietveld refinements, applying corrections for preferred orientation, to verify that the intensities were reliable. Once the phases involved were defined, lattice parameters were obtained by Le Bail refinements.

The calorimetric analysis was carried out on fully annealed samples using a differential scanning calorimeter (mod. SSC 5200 by Seiko Instruments, Chiba, Japan) at a rate of 10 °C/min; heating/cooling scans were performed at temperatures ranging from −100 °C to 450 °C for a number of 10 thermal cycles. Characteristic temperatures and the corresponding transformation heats were evaluated for both direct/reverse transformations upon cooling/heating scans, respectively. Microhardness measurements were performed on the cross section of the samples using a Future Tech Corporation FM-700 hardness tester (Future-Tech Corp FM-700, Tokyo, Japan), applying a 200 g load for 15 s. Compressive tests were conducted at room temperature by means of an MTS 2/M machine (MTS Systems Corporation, Eden Prairie, MN, USA), equipped with extensometer, at a strain rate of 0.015 min^{-1}; five complete cycles, loading and unloading up to 1% in strain, were carried out on cylindrical samples (3 mm in diameter and 8 mm in height).

3. Results and Discussion

SEM images of the samples and related EDS analyses are shown in Figure 1 and Table 1, respectively. The binary $Cu_{50}Zr_{50}$ alloy (see Figure 1a) appears to consist of a CuZr matrix containing the secondary phase $Cu_{10}Zr_7$. This result is consistent with previous work [7,9,14,24].

Table 1. EDS measurements of the CuZr based sample, performed on the sites depicted in Figure 5.

Composition	Phases	Measured Compositions (at. %)
$Cu_{50}Zr_{50}$ (see Figure 1a)	(1) CuZr	$Cu_{47}Zr_{53}$
	(2) $Cu_{10}Zr_7$	$Cu_{61}Zr_{39}$
$Cu_{45}Zr_{50}Al_5$ (see Figure 1b)	(1) CuZr	$Al_{5.5}Cu_{46}Zr_{48.5}$
	(2) τ5	$Al_{18}Cu_{42}Zr_{40}$
	(3) $O_x(CuZr_2)$	$Al_{12}Cu_{31}Zr_{46}O_{11}$
$Cu_{40}Zr_{50}Al_{10}$ (see Figure 1c)	(1) CuZr	$Al_6Cu_{45}Zr_{49}$
	(2) τ5	$Al_{21}Cu_{41}Zr_{38}$
	(3) $CuZr_2$	$Al_4Cu_{35}Zr_{61}$
	(4) $O_x(CuZr_2)$	$Al_{16}Cu_{28}Zr_{44}O_{12}$
$Cu_{35}Zr_{50}Al_{15}$ (see Figure 1d)	(1) CuZr	$Al_{5.5}Cu_{46.5}Zr_{48}$
	(2) τ5	$Al_{23}Cu_{40}Zr_{37}$
	(3) $CuZr_2$	$Al_5Cu_{37}Zr_{58}$
	(4) $O_x(CuZr_2)$	$Al_{17}Cu_{30}Zr_{46}O_7$

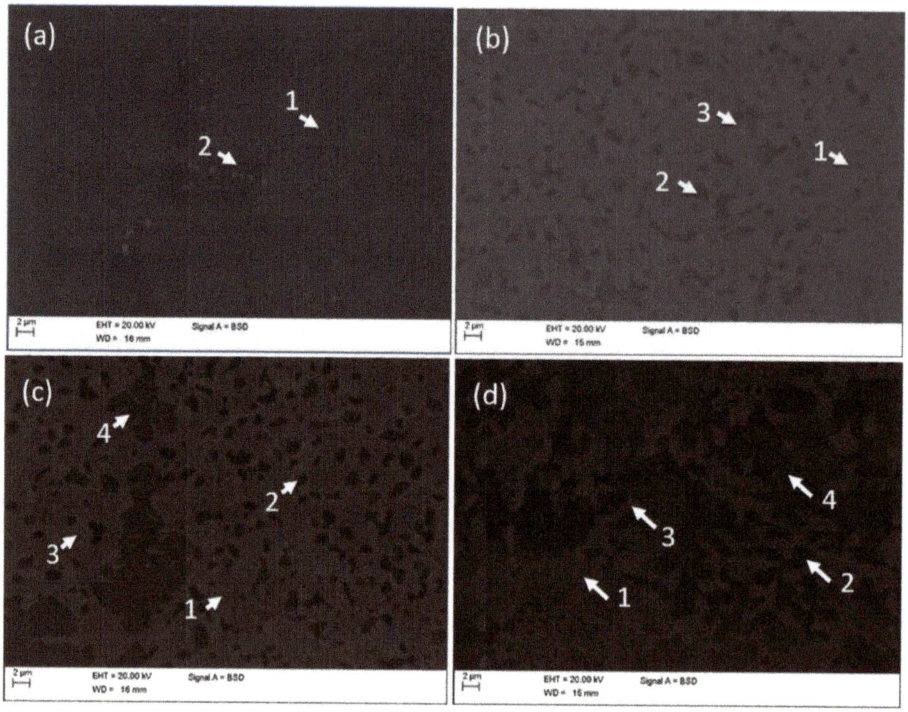

Figure 1. BSE-SEM images of $Cu_{50}Zr_{50}$ after MT stabilization (**a**) and $Cu_{(50-x)}Zr_{50}Al_x$ with x = 5, 10, 15 at. % (**b–d**) before thermal cycling, respectively.

Figure 1b shows the microstructure of $Cu_{45}Zr_{50}Al_5$. A predominance of the CuZr matrix (light colored area) containing small traces of a ternary phase (dark areas) and some cross-shaped oxides (grey areas) can be seen. As listed in Table 1, the ternary phase exhibits an average composition of Cu_2Zr_2Al. A very similar secondary phase, in terms of morphology and elemental composition, was observed by Meng et al. [20] in samples annealed at the same temperature as in the present investigation. After comparison with XRD analysis, this secondary phase was identified as τ5, although the compositional stability range of the τ5 phase was previously proposed to be limited to $ZrCu_xAl_{2-x}$ with x ranging from 0.34 to 0.95 [24–26].

A further increase in Al content (10 at. %) induced the formation of $CuZr_2$ and an increase in the number of secondary phases (see Figure 1c). Numerous dark regions of τ5, with average composition $Zr_{40}Cu_{42}Al_{18}$ (at. %), are dispersed into the CuZr matrix. Finally, cross-like oxides, characterized by a composition similar to the ones found in $Cu_{45}Zr_{50}Al_5$, were also detected. The brighter areas surrounding τ5 and oxide particles were found to exhibit a composition lying in the range of the $CuZr_2$ phase. In the sample containing 15% Al (see Figure 1d), the high amount of aluminum induced a strong increase in the content of the τ5 phase (dark areas). A high amount of $CuZr_2$, some of which oxidized, was also found. The formation of the $CuZr_2$ oxides, already reported in Cu-Zr system [6], is considered to take place during melting, according to the high oxygen reactivity of Zr, which is similar to that of Ti. It is well known that the oxides inhibit the shape memory behavior in quasi-equiatomic NiTi alloys; in the present system, the influence of these oxides on the MT, although not clear, can probably be considered similar to that of equiatomic NiTi alloys.

A remark is due on the distribution of Al within the different phases. Concerning the CuZr matrix, EDS analysis revealed that the Al content, present as solid solution in the CuZr matrix, is constant: an average content of 5.5%, 6.0%, and 5.5% was found in

$Cu_{(50-x)}Zr_{50}Al_x$ with x = 5, 10, 15 at. %, respectively. These values match those reported by Meng et al. [20] well, who stated that the solubility limit of Al in the CuZr phase is about 5 at. %. Since the CuZr phase is always found to be rich in Zr, it is assumed that Al might preferentially substitute for Cu rather than for Zr. On the other hand, the correct $CuZr_2$ stoichiometry can be guaranteed only by considering that the doping Al substitutes for Zr. This is particularly true as far as the related oxide, which is well known to form cross-like sections with a $CuZr_2O_{0.5}$ base composition, is concerned. In this case, the oxidized areas in samples $Cu_{40}Zr_{50}Al_{10}$ and $Cu_{35}Zr_{50}Al_{15}$ were found to contain a high amount of aluminum, up to 20 at. %. The metallic radius of Al (143 pm), being exactly intermediate between those of Cu and Zr (128 pm and 160 pm [24], respectively) is likely to lie at the basis of this apparently controversial behavior.

Figure 2 displays the experimental XRD patterns of binary and ternary alloys, focusing on the 20 to 50 degrees region, where most of the main peaks occur. Different labels were employed to mark the peaks of the main phases. As reported by Biffi et al. [18], before any thermal cycles the equiatomic binary alloy is composed of the two monoclinic martensites. In addition to these, $Cu_{10}Zr_7$ and $CuZr_2$ secondary phases are present in the Cu/Zr ratio close to 1:1. The blurred features of the pattern confirm that most of the sample is composed of martensites, the contributions of which, being quite close to each other, were not resolved in the indexing, as is generally observed for CuZr compounds [15].

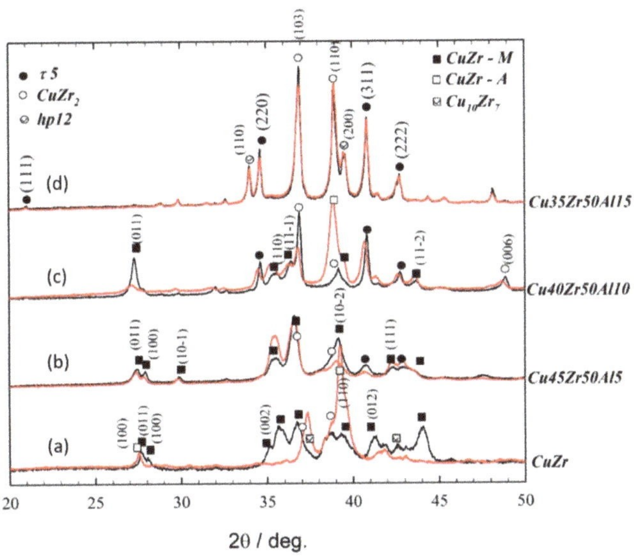

Figure 2. Section of experimental patterns of binary equiatomic (**a**) and ternary alloys (**b–d**) at room temperature, before (black curve) and after (red curve) the tenth thermal cycle. The reflections of the τ5 phase are explicitly indicated.

The addition of Al progressively consumes CuZr, in agreement with the decreasing heat exchange from DSC, leading to the preferential growth of two different phases: $CuZr_2$ and τ5. The latter phase can be clearly distinguished since the intensity of its main four reflections, well resolved from other contributions, steeply increases with Al amount. The position and relative intensities of the peaks are consistent with a phase having space group Fd-3m and lattice parameter 7.326 Å. The compositional range of stability of the τ5 phase has then to be reconsidered from that originally proposed in [14,15]. Besides the τ5 phase, Al addition increases the proportion of the $CuZr_2$ phase as well.

The lattice parameters for the main phases of interest are reported in Table 2. The B19′ martensite is actually composed of two phases, however, as the second phase is a superstructure of the former and most of the reflections are shared, here we consider only the $P2_1/m$ for the sake of simplicity. While the absolute values might be an average of those of the two phases, they can provide information about the lattice response to Al incorporation. The presence of Al shrinks the CuZr lattice parameters, especially axes a and b. This is consistent with the slight enrichment of Cu in place of Zr, making Al substituting more likely for the latter. Similarly, the incorporation of Al within $CuZr_2$ is confirmed by the evident contraction of the lattice parameter c, from 11.20 Å for the binary alloy down to 11.05 Å, which is also consistent with the progressive enrichment of Cu (and depletion of Zr) observed via SEM-EDS. On the other hand, the lattice parameter of the τ5 phase is nearly invariant across the different compositions.

Table 2. Lattice parameters (Å and °) of martensite CuZr ($P2_1/m$) and $CuZr_2$ as a function of Al content. Uncertainties are estimated to be 0.005 Å and 0.2°.

Composition	CuZr				$CuZr_2$	
	a	b	c	β	a	c
$Cu_{50}Zr_{50}$	3.253	4.394	4.964	105.2	3.253	11.200
$Cu_{45}Zr_{50}Al_5$	3.171	4.433	4.910	104.6	3.219	11.148
$Cu_{40}Zr_{50}Al_{10}$	3.144	4.244	4.920	104.9	3.201	11.166
$Cu_{35}Zr_{50}Al_{15}$	-	-	-	-	3.160	11.045

SEM-EDS analysis indicates that Al shows the tendency of substituting either for Cu or for Zr, depending on the phase of interest. Therefore, increasing the Al amount in $Zr_{50}Cu_{50-x}Al_x$ increases the relative amount of Zr vs. Cu, thus moving towards the Zr-rich part of the phase diagram, i.e., stabilizing the $CuZr_2$ phase. Indeed, we observed the preferential formation of $CuZr_2$ rather than $Cu_{10}Zr_7$ as a secondary phase [27].

Finally, the XRD pattern related to the Al addition of 15 at. % clearly shows the presence of contributions other than CuZr, $CuZr_2$, and τ5. Beside a minor contribution of the τ10 phase, two big reflections at ~34.0° and 39.5° were tentatively indexed with a hexagonal phase (hp12 in Figure 5) related to that of $ZrAl_2$. It cannot be excluded that this phase may be related to the Al-rich oxide observed by SEM-EDS. The different elemental composition between binary and ternary alloys does not only define the nature and amount of the phases; it also drastically affects stability against thermal cycling.

Figure 2 compares the experimental XRD patterns of binary CuZr (see Figure 3a) and ternary CuZrAl alloys with 5, 10 and 15 at. % Al (see Figure 3b–d) before and after 10 thermal cycles across MT. The pattern of the binary alloy is definitely altered upon cycling, owing to both the transformation from fully martensitic to fully austenitic phases and to the increased amount of secondary $Cu_{10}Zr_7$ and $CuZr_2$ phases. It is worth noting that the cycling was performed in the diffractometer, hence the sample position was not altered and all modifications observed in the XRD patterns are ascribable to the structure evolution only. On the other hand, ternary alloys are not significantly modified upon cycling. The two patterns of sample with 15 at. % Al are very similar, testifying that the same phase composition and relative concentration are maintained upon cycling. Only a slight increase in peak broadening was observed, suggesting an accumulation of defects or residual stresses.

DSC scans of stabilized $Cu_{50}Zr_{50}$ and $Cu_{(50-x)}Zr_{50}Al_x$ alloys in fully annealed condition are shown in Figure 3. As previously reported by the authors [9,10], the binary alloy exhibits unstable behavior. The MT temperatures are initially above 100 °C. Upon thermal cycling, they are shifted down to lower temperatures (martensite start temperature, M_s = 150 °C at the first cycle, M_s = −20 °C at the 8th cycle). After the 8th cycle no variation in DSC signals is detected, as depicted in Figure 3a, and the MT shows stable behavior below 0 °C. The stabilized transformation is also characterized by a much narrower thermal hysteresis (less than 10 °C).

It is evident that 5 at. % Al addition promotes MT thermal cycling stability (see Figure 3b). Furthermore, the Al addition suddenly decreases MT transformation temperatures from the very first thermal cycle, as depicted in the DSC scan magnification of Figure 3b–d where a stable MT peak with M_s of −54 °C is visible. The increasing of Al content up to 15 at. % does not affect the thermal stability of the MT but it reduces transformation temperatures as well as DSC signals, as reported in Figure 3c,d, for 10 and 15 at. % of Al, respectively. As displayed in Figure 4a, the addition of 5 at. % Al induces an evident decrease in the transformation temperatures, whereas the further increase towards 10 at. % and 15 at. % of Al causes a more progressive depression of the characteristic temperatures. The enthalpy values associated with forward and reverse phase transformations are reported in Figure 4b as a function of the Al content: the increase of the Al content decreases both enthalpies from about 3 J/g, for the binary alloy, down to 0.5 J/g, for the ternary alloy with 15 at. % Al content. This may be explained by considering that the amount of the phase involved in the MT, i.e., the Al-free Cu-Zr, decreases, as seen for the binary CuZr alloys too [6]. In fact, in this work it was found that the equiatomic phase is associated with the MT, and a shift of the chemistry of the alloy away from the 1:1 Cu to Zr ratio indicates a decrease in the enthalpy involved during the MT. The DSC results are only in partial agreement with the work done by Meng et al. [20]: the shape of the peaks of the MT for all investigated Al contents is slightly different between the literature and the current work. Another significant difference is that the MT occurred at temperatures below 0 °C (see Figure 4), while Meng et al. indicated that the MT occurred at higher temperatures [20]. This was probably due to the grain size and the specific melting system adopted for the sample preparation. Moreover, the current heat treatment was carried out for 3 h while in Meng's work it took 72 h.

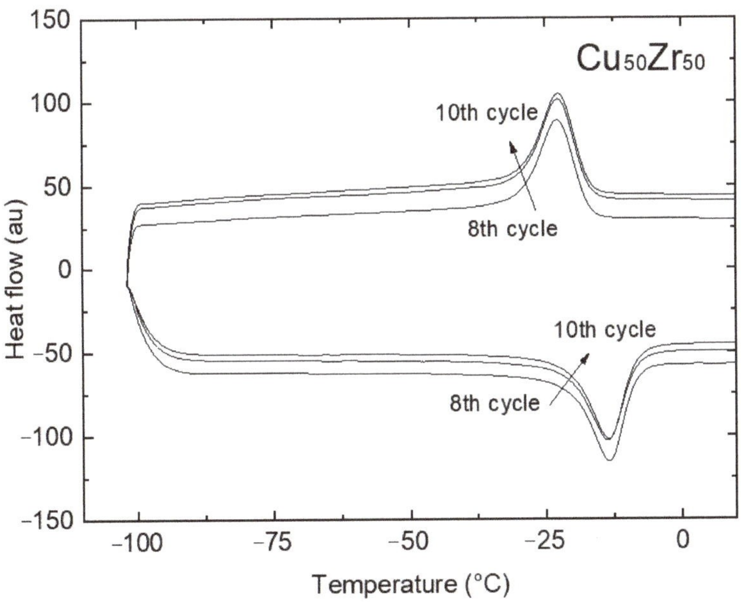

(a)

Figure 3. *Cont.*

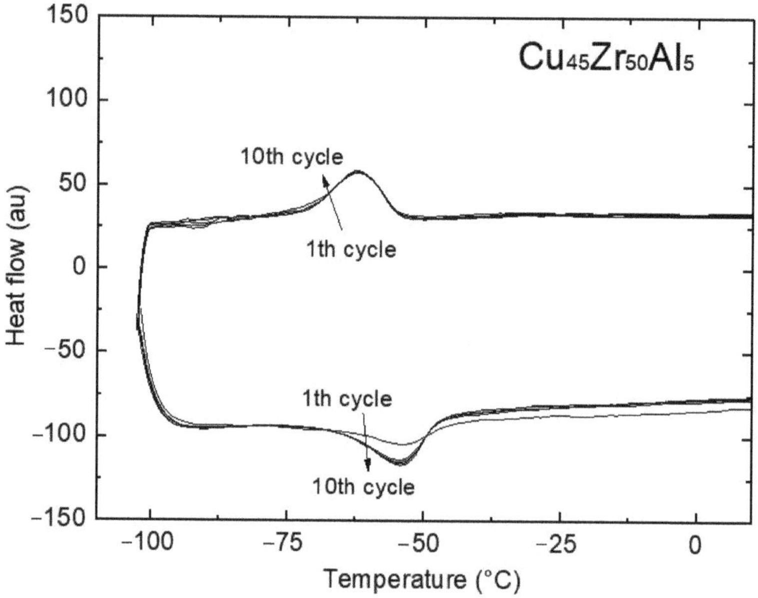

(**b**)

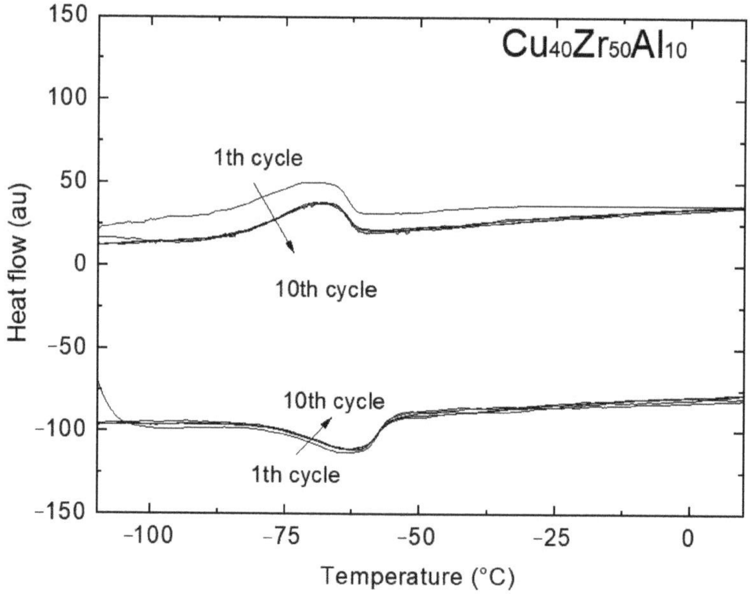

(**c**)

Figure 3. *Cont.*

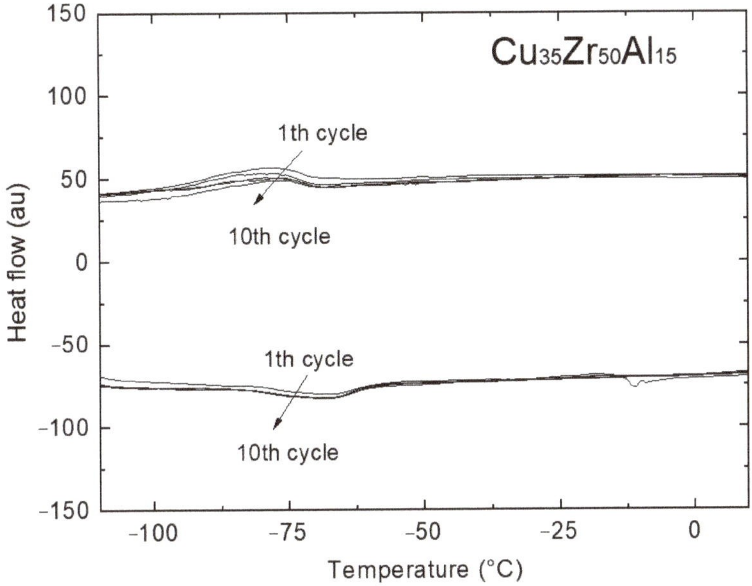

(**d**)

Figure 3. DSC scans of CuZr based alloys: MT after stabilization of $Cu_{50}Zr_{50}$ alloy (**a**); thermal cycling of $Cu_{(50-x)}Zr_{50}Al_x$ alloys with Al = 5, 10, 15 at. % (**b–d**), respectively.

Additionally, the thermal hysteresis increases when moving from the equiatomic to the ternary alloys, and it is directly proportional to the Al content, as previously indicated. The transformation enthalpies, which are related to the CuZr phase transformation [7], decrease as Al content increases. This might suggest a lowering of the amount of CuZr phase involved in the MT, as below reported. As shown in Figure 5, the trend of microhardness of the investigated alloys indicates a softening moving from the binary $Cu_{50}Zr_{50}$, having an average hardness of 550 HV, to the ternary alloy with the lowest Al content, having an average hardness of 450 HV. Increasing the Al content from 5 at. % to 10 at. % does not modify the hardness values while moving up to 15 at. % of aluminum induces an increase in the mechanical properties. A similar trend in $(CuZr_2)_{100-x}Al_x$ was observed by Qiu et al. [18]. This trend appears to be in good agreement with the mechanical response of the alloys tested in compression during five complete loading/unloading cycles.

Figure 6 shows the stress/strain curves, obtained at room temperature during quasi-static compression testing, for the stabilized equiatomic CuZr alloy and for the three ternary alloys. Uniaxial compression testing was selected for studying the mechanical response of these alloys, as it has been widely used to study the mechanical properties of materials having a lack of tensile ductility. The trend of the maximum stress, recoverable, and residual strains during mechanical cycling as a function of the Al content are reported in Figure 7.

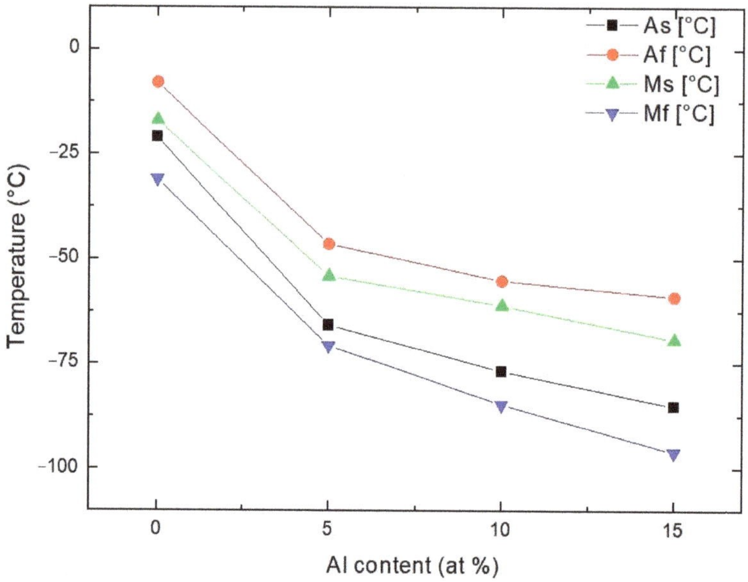

(**a**)

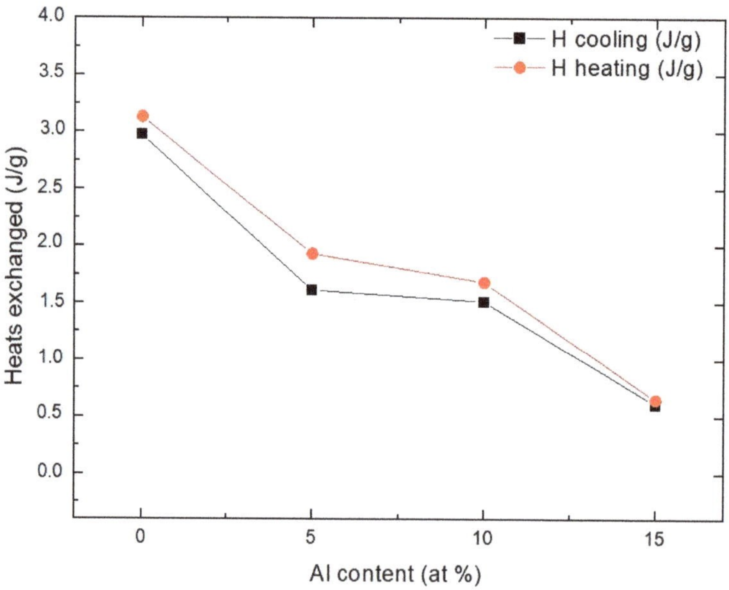

(**b**)

Figure 4. Transformation temperatures (**a**) and heat exchanged (**b**) of $Cu_{(50-x)}Zr_{50}Al_x$ with x = 0, 5, 10, 15 at. % after MT stabilization.

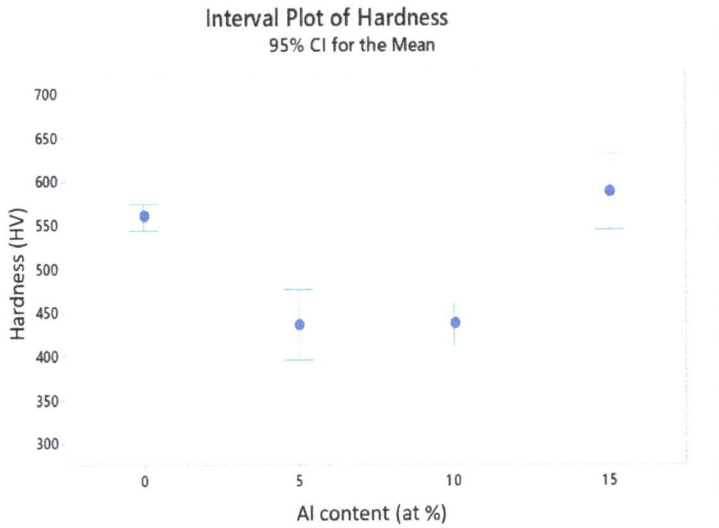

Figure 5. Microhardness as a function of the Al content.

In details, the thermally stabilized binary CuZr alloy shows perfect linear behavior during the first cycle, while the second one exhibits a superelastic behavior with a more than intensive linear strain hardening effect, which is correlated with the high temperature phase (see DSC scan of Figure 3a). This response remains present up to the fifth cycle, during which the failure occurred before reaching the maximum stress during the loading. The addition of Al in the range between 5% and 10% provokes an increase in the values of the recoverable strain during each cycle and the failure cannot be reached within the fifth cycle. The loading and unloading slopes are much lower for these compositions of the ternary system, if compared with the binary alloy. On the contrary, higher Al content clearly induces a brittle response (see Figure 6d), which is supported by an increase in brittleness. The alloy fragility could be correlated with the accumulation of defects or residual stresses, previously detected by the XRD analysis. In fact, the failure of the last composition was found during the loading of the fourth cycle, therefore corresponding to the minor number of mechanical cycles.

As shown in Figure 7a, the maximum stress can be correlated with the Al content similarly to the microhardness values: this behavior indicates that the effect of Al addition on the mechanical properties can be well estimated. As the binary alloy has the highest recoverable strain, its failure occurs in the fifth cycle; similar to the behavior of the alloy with the highest Al content. On the contrary, 5at. % and 10at. % of Al addition promote the best recoverable performances during the mechanical cycling without reaching the failure of the samples up to 2% in correspondence of 0.5% of residual strain (see Figure 7b). The binary CuZr was characterized by the highest value of residual strain, up to 1%. All the ternary alloys were associated with a decrease of the residual strain, indicating an improvement of the elastic behavior, as shown in Figure 7c.

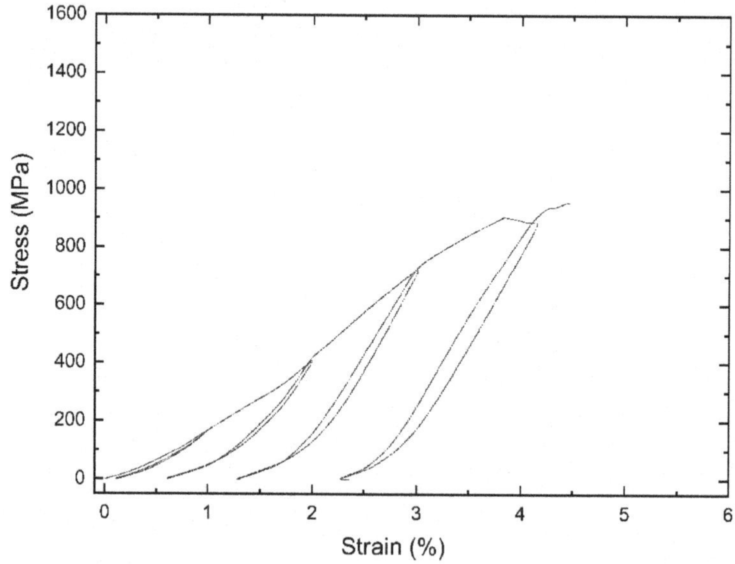

(a)

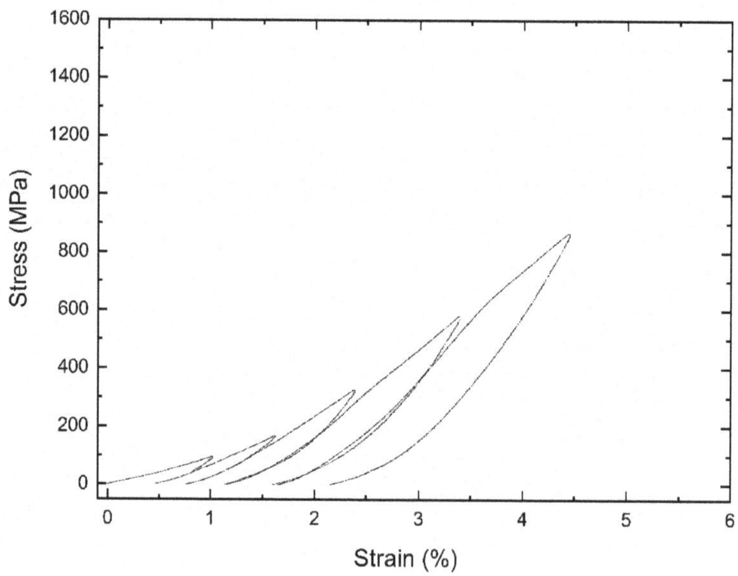

(b)

Figure 6. *Cont.*

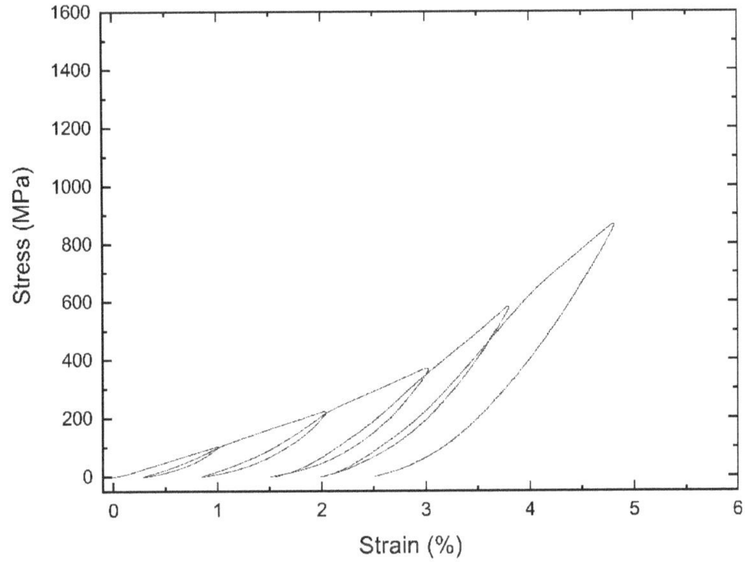

(**c**)

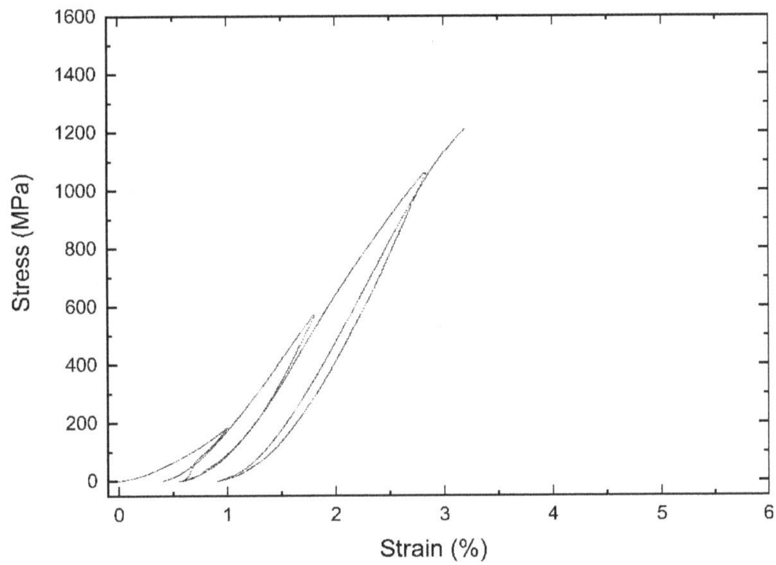

(**d**)

Figure 6. Compression behavior of CuZr based alloys, tested at room temperature: stress–strain after stabilization of $Cu_{50}Zr_{50}$ alloy (**a**); and $Cu_{(50-x)}Zr_{50}Al_x$ alloys with Al = 5, 10, 15 at. % (**b**–**d**), respectively. The stresses and strains are compressive.

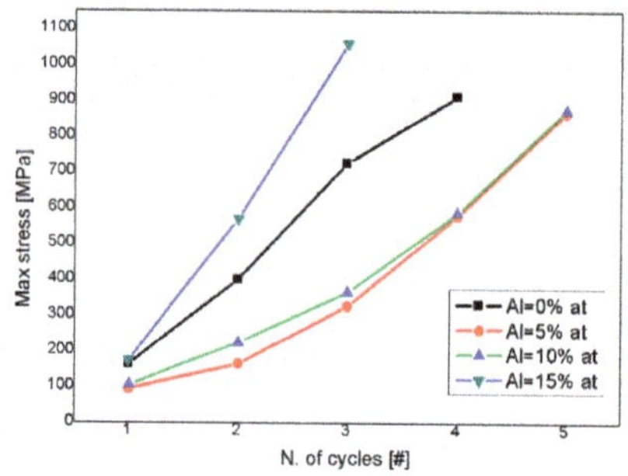

(**a**)

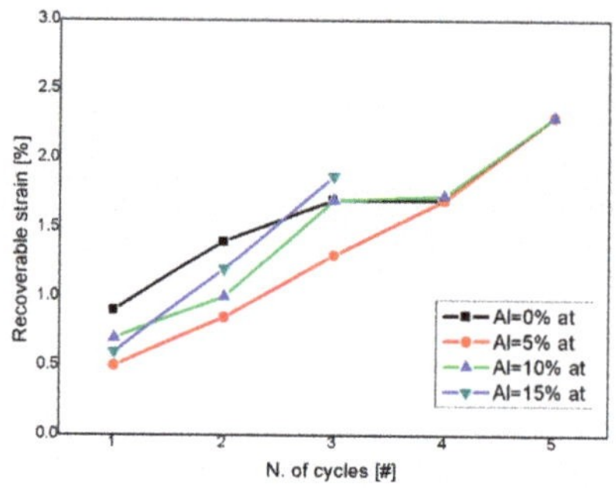

(**b**)

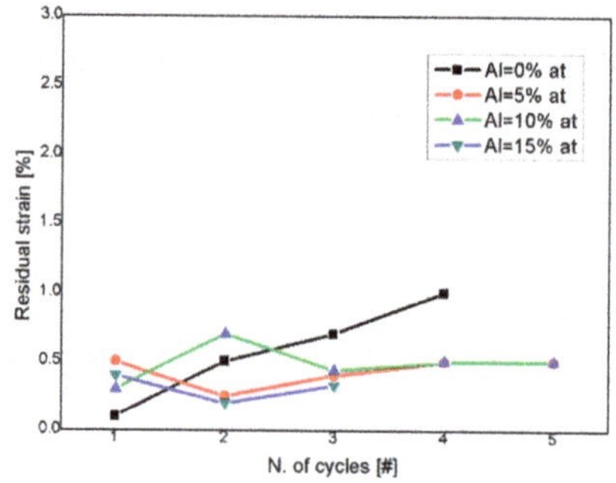

(**c**)

Figure 7. Trend of maximum stress (**a**), recoverable (**b**), and residual strain (**c**) of the $Cu_{(50-x)}Zr_{50}Al_x$ alloys with Al = 0, 5, 10, 15 at. %. The stresses and strains are compressive.

4. Conclusions

The present work explores the possibility of alloying a binary $Cu_{50}Zr_{50}$ alloy by adding different amounts of Al (5, 10, and 15 at. %) in place of Cu. The presence of aluminum causes a reduction in the proportion of the CuZr phase and the formation of secondary phases (τ5, $CuZr_2$). These microstructural changes strongly influenced the characteristics of MT: while in the binary alloy the transformation is stabilized only after eight thermal cycles, samples containing aluminum exhibit stable MT ab initio. Transformation temperatures and enthalpies get lower as the aluminum content is increased: this effect does not permit any possibilities in the use of these alloys as HTMAs. The Al alloying, in the range from 5% up to 10%, made evident a pseudoelastic behavior under compression, while higher Al content induced high brittleness, in a similar manner to the binary CuZr alloy after MT stabilization.

It can be concluded that Al addition to equiatomic alloy, although it stabilizes the martensitic transformation, does not allow for the use of the alloy as an HTSMA. Further studies on different alloying elements focused on finding a proper balance between thermal cycling stability and suitable high transformation temperatures are required before using a CuZr based system for the development of HTSMA for industrial applications.

Author Contributions: Conceptualization, C.A.B. and A.T.; methodology, C.A.B.; validation, all authors.; formal analysis, C.A.B., J.F. and M.C.; investigation, C.A.B., J.F. and M.C.; resources, A.T.; writing—original draft preparation, C.A.B., J.F. and M.C.; writing—review and editing, all authors; visualization, C.A.B.; supervision, A.T.; project administration, A.T.; funding acquisition, A.T. All authors have read and agreed to the published version of the manuscript.

Funding: This research was funded through Accordo Quadro between CNR and Regione Lombardia, grant number 3866.

Data Availability Statement: Data available on request.

Acknowledgments: The work was developed within the framework of 2° Accordo Quadro CNR/Regione Lombardia. The authors would like to thank Marco Pini, Nicola Bennato, and Giordano Carcano of CNR ICMATE, Lecco Unit, for their technical assistance.

Conflicts of Interest: The authors declare no conflict of interest.

References

1. Otsuka, K.; Wayman, C.M. *Shape Memory Materials*; Cambridge University Press: New York, NY, USA, 1998.
2. Hodgson, D.E. Fabrication, Heat Treatment and Joining of Nitinol Components. In *SMST-2000: Proceedings of the International Conference on Shape Memory and Superelastic Technologies*; ACM: New York, NY, USA, 2001; pp. 11–24.
3. Koval, Y.; Firstov, G.; Kotko, A. Martensitic transformation and shape memory effect in ZrCu intermetallic compound. *Scr. Met. Mater.* **1992**, *27*, 1611–1616. [CrossRef]
4. Schryvers, S.D. TEM investigation of the microstructure and defects of CuZr martensite. Part I: Morphology and twin systems. *Acta Mater.* **1998**, *46*, 1165–1175.
5. Seo, J.W.; Schryvers, D. EM investigation of the microstructure and defects of CuZr martensite. Part II: Planar defects. *Acta Mater.* **1998**, *46*, 1177–1183.
6. Carvalho, E.M.; Harris, I.R. Constitutional and structural studies of the intermetallic phase, ZrCu. *J. Mater. Sci.* **1980**, *15*, 1224–1230. [CrossRef]
7. Biffi, C.; Figini, A.; Tuissi, A. Influence of compositional ratio on microstructure and martensitic transformation of CuZr shape memory alloys. *Intermetallics* **2014**, *46*, 4–11. [CrossRef]
8. Firstov, G.S.; Van Humbeeck, J.; Koval, Y.N. Peculiarities of the martensitic transformation in ZrCu intermetallic compound—potential high temperature SMA. *J. Phys. Colloq.* **2001**, *11*, 481–486. [CrossRef]
9. Biffi, C.; Coduri, M.; Yoshida, H.; Soejima, Y.; Nishida, M.; Tuissi, A. The effect of thermal cycling on the martensitic transformation in equiatomic CuZr shape memory alloy. *J. Alloy. Compd.* **2015**, *653*, 591–595. [CrossRef]
10. Hisada, S.; Matsuda, M.; Nishida, M.; Biffi, C.A.; Tuissi, A. Microstructure and Martensitic Transformation Behavior in Thermal Cycled Equiatomic CuZr Shape Memory Alloy. *Metals* **2019**, *9*, 580. [CrossRef]
11. Song, K.; Pauly, S.; Zhang, Y.; Gargarella, P.; Li, R.; Barekar, N.; Kühn, U.; Stoica, M.; Eckert, J. Strategy for pinpointing the formation of B2 CuZr in metastable CuZr-based shape memory alloys. *Acta Mater.* **2011**, *59*, 6620–6630. [CrossRef]
12. Firstov, G.S.; Van Humbeeck, J.; Koval, Y.N. High-temperature shape memory alloys: Some recent developments. *Mater. Sci. Eng.* **2004**, *378*, 2–10. [CrossRef]

13. Firstov, G.S.; van Humbeeck, J.; Koval, Y.N. Comparison of high temperature shape memory behaviour for ZrCu-based, Ti–Ni–Z and Ti–Ni–Hf alloys. *Scripta Mater.* **2004**, *50*, 243–248. [CrossRef]
14. Biffi, C.A.; Albisetti, A.F.; Tuissi, A. CuZr Based Shape Memory Alloys: Effect of Cr and Co on the Martensitic Transformatio *Mater. Sci. Forum* **2013**, *738–739*, 167–171. [CrossRef]
15. Koval, Y.; Firstov, G.; Delaey, L.; Humbeeck, J. The influence of Ni and Ti on the martensitic transformation and shape memor effect of the intermetallic compound ZrCu. *Scr. Metall. Mater.* **1994**, *31*, 799–802. [CrossRef]
16. Biffi, C.A.; Tuissi, A. Hot Workability of Cu-Zr Based Shape Memory Alloys for Potential High Temperature Applications. *Mater. Eng. Perform.* **2014**, *23–27*, 2379. [CrossRef]
17. Gao, W.; Yi, X.; Song, G.; Wang, Z.; Meng, X. $Zr_{50}Cu_{25}Ni_{7.5}Co_{17.5}$ high-temperature shape memory alloy with excellent therma stability and large recovery strain, and the associated microstructural deformation mechanism. *Mater. Des.* **2020**, *196*, 10910 [CrossRef]
18. Qiu, F.; Shen, P.; Liu, T.; Lin, Q.; Jiang, Q. Electronic structure and phase stability during martensitic transformation in Al-dope ZrCu intermetallics. *J. Alloy. Compd.* **2010**, *491*, 354–358. [CrossRef]
19. Qiu, F.; Wang, H.; Liu, T.; Jiang, Q. Influence of Al content on the microstructure and mechanical property of the $(Zr_2Cu)_{100-x}Al$ alloys. *J. Alloy. Compd.* **2009**, *468*, 195–199. [CrossRef]
20. Meng, F.Q.; Tsuchiya, K.; Yin, F.X.; Li, S.; Yokoyama, Y. Influence of Al content on martensitic transformation behaviour i $Zr_{50}Cu_{50-x}Al_x$. *J. Alloy. Compd.* **2012**, *522*, 136–140. [CrossRef]
21. Gao, W.; Meng, X.; Cai, W.; Zhao, L. ffects of Co and Al addition on martensitic transformation and microstructure in ZrCu-base shape memory alloys. *Trans. Nonferrous Met. Soc. China* **2015**, *25*, 850–855. [CrossRef]
22. Barekar, N.; Gargarella, P.; Song, K.; Pauly, S.; Kühn, U.; Eckert, J. Effect of Al and Ag addition on phase formation, therma stability, and mechanical properties of Cu–Zr-based bulk metallic glasses. *J. Mater. Res.* **2011**, *26*, 14. [CrossRef]
23. Kwon, O.J.; Kim, Y.C.; Kim, K.B.; Lee, Y.K.; Fleury, E. Formation of amorphous phase in the binary Cu−Zr alloy system. *Me Mater. Int.* **2006**, *12*, 207–212. [CrossRef]
24. Massalski, T.B.; Okamoto, H.; Subramanian, P.R.; Kacprzak, L. *Binary Alloy Phase*, 2nd ed.; ASM International: Materials Park OH, USA, 1990.
25. Raghavan, V. Al-Cu-Zr (Aluminum-Copper-Zirconium). *J. Phase Equilibria Diffus.* **2011**, *32*, 452–454. [CrossRef]
26. Wang, C.P.; Tu, S.B.; Yu, Y.; Han, J.J.; Liu, X.J. Experimental investigation of phase equilibria in the Zr-Cu-Al system. *Intermetallics* **2012**, *31*, 1–8. [CrossRef]
27. Albisetti, A.F.; Biffi, C.; Tuissi, A. Synthesis and structural analysis of $Cu_{10}Zr_7$. *J. Alloy. Compd.* **2012**, *544*, 42–45. [CrossRef]

metals MDPI

Article

Martensitic Transformation and Metamagnetic Transition in Co-V-(Si, Al) Heusler Alloys

Kousuke Nakamura [1], Atsushi Miyake [2], Xiao Xu [1,*], Toshihiro Omori [1], Masashi Tokunaga [2] and Ryosuke Kainuma [1]

[1] Department of Materials Science, Graduate School of Engineering, Tohoku University, Sendai 980-8579, Japan; kousuke.nakamura.s3@dc.tohoku.ac.jp (K.N.); omori@material.tohoku.ac.jp (T.O.); kainuma@material.tohoku.ac.jp (R.K.)
[2] Institute for Solid State Physics, The University of Tokyo, Kashiwa, Chiba 277-8581, Japan; miyake@issp.u-tokyo.ac.jp (A.M.); tokunaga@issp.u-tokyo.ac.jp (M.T.)
* Correspondence: xu@material.tohoku.ac.jp; Tel.: +81-22-795-7323

Abstract: This study investigates the crystal structure, martensitic transformation behavior, magnetic properties, and magnetic-field-induced reverse martensitic transformation of $Co_{64}V_{15}(Si_{21-x}Al_x)$ alloys. It was found that by increasing the Al composition, the microstructure changes from the martensite phase to the parent phase. The crystal structures of the martensite and parent phases were determined as $D0_{22}$ and $L2_1$, respectively. Thermoanalysis and thermomagnetization measurements were used to determine the martensitic transformation and Curie temperatures. Both the ferromagnetic state of the parent phase and that of the martensite phase were observed. With the increasing Al contents, the martensitic transformation temperatures decrease, whereas the Curie temperatures of both the martensite and parent phases increase. The spontaneous magnetization and its composition dependence were also determined. The magnetic-field-induced reverse martensitic transformation of a $Co_{64}V_{15}Si_7Al_{14}$ alloy under pulsed high magnetic fields was observed. Moreover, using the results of the DSC measurements and the pulsed high magnetization measurements, the temperature dependence of the transformation entropy change of the Co-V-Si-Al alloys was estimated.

Keywords: co-based Heusler alloy; martensitic transformation; metamagnetic shape memory alloy; phase diagram; magnetic-field-induced transition

Citation: Nakamura, K.; Miyake, A.; Xu, X.; Omori, T.; Tokunaga, M.; Kainuma, R. Martensitic Transformation and Metamagnetic Transition in Co-V-(Si, Al) Heusler Alloys. *Metals* **2021**, *11*, 226. https://doi.org/10.3390/met11020226

Academic Editor: Gabriel A. Lopez
Received: 14 December 2020
Accepted: 21 January 2021
Published: 28 January 2021

1. Introduction

Shape memory alloys showing both shape memory effects and/or superelasticity are important functional materials, and many shape memory alloys have been reported, such as the NiTi [1], Cu-based [2], Fe-based [3], and Mg-based [4] alloy systems. Recently, ferromagnetic shape memory alloys have received extensive attention because the magnetic-field-induced strain by variant rearrangement [5] and the magnetic-field-induced reverse martensitic transformation [6] have been found in Ni-Mn-X (X = In, Sn, Sb [7], and Ga [5]) Heusler alloys. In contrast, Co-based Heusler alloys are attractive in the field of spintronics owing to their half-metallic behavior [8]. Owing to the high phase stability of the Heusler phases, the martensitic transformation was not observed in the Co-based Heusler alloys except for Co_2NbSn, which does not exhibit half-metallic behavior [9]. However, recently, martensitic transformations in Co-based Heusler alloys, such as Co-Cr-Ga-Si [10], Co-Cr-Al-Si [11], Co-V-Ga [12], Co-V-Si [13], and Co-V-Al [14], have been reported. Among them, the magnetic-field-induced phase transition was realized in Co-Cr-Ga-Si, Co-Cr-Al-Si, and Co-V-Ga alloys [15–17]. Furthermore, the Co-Cr-Ga-Si alloys show anomalous behavior, called reentrant martensitic transformation, which is a transformation from the paramagnetic martensite phase to the reentrant ferromagnetic parent phase, by cooling in addition to the normal martensitic transformation from the paramagnetic parent phase to the paramagnetic martensite phase [10]. Using the reentrant martensitic transformation, a cooling-induced shape memory effect was realized. This paper focuses on the CoV-based alloy system. In the Co-V-Si system, a martensitic transformation from $L2_1$ to $D0_{22}$ structures with decreasing temperature is found to occur at a high temperature of approximately 700 °C; thus, the use of a high-temperature shape memory alloy is considered [13]. However, the temperature is so high that a phase decomposition from $L2_1$ to A12 arises. Consequently, the forward martensitic transformation during cooling after heating to the parent phase cannot be detected owing to the diffusional transformation, and the sample degenerates only after one test of the shape memory effect for the as-quenched sample. To solve this problem, studies have been conducted by adding group III elements to lower the martensitic transformation temperature. For example, Ga-doped Co-V-Si-Ga alloys have been reported [18]. The addition of Ga has been found to be useful for decreasing the martensitic transformation temperature and improving the thermal stability, although also increasing the cost. Additionally, an increasing Ga composition improved the recovered strain of the shape memory effect [15]. Very recently, a good thermal stability of more than 200 thermal cycles from room temperature to 850 °C was reported in a Co-V-Si-Al alloy with a small concentration of Al [19]. Therefore, the Co-V-Si-Al system is expected to be an inexpensive solution for application as high-temperature shape memory alloys. However, systematic studies on quaternary alloys have not been conducted. In this study, the crystal structures, the martensitic transformation behavior, and magnetic properties of the $Co_{64}V_{15}(Si_{21-x}Al_x)$ alloys were investigated, and a pseudo-binary magnetic phase diagram was determined. Furthermore, by the application of a pulsed high magnetic field, the magnetic-field-induced reverse martensitic transformation was realized, which also opens the possibility of its use as a multifunctional magnetic material for this new alloy system. Moreover, the temperature dependence of the transformation entropy, which had only been estimated for the Co-Cr-Ga-Si and Co-Cr-Al-Si alloys in the Co-based Heusler alloys [15,20], was further investigate by using the results of differential scanning calorimeter (DSC) measurements and magnetization measurements in the pulsed fields.

2. Materials and Methods

$Co_{64}V_{15}(Si_{21-x}Al_x)$ ($0 \le x \le 21$) alloys were prepared from high-purity Co (99.9%), V (99.7%), Si (99.999%), and Al (99.99%) by arc melting in an argon atmosphere. The alloys were sealed into quartz tubes and solution-heat-treated at 1473 K for 24 h and then quenched in ice water by breaking the tubes. The microstructures were observed using a scanning electron microscope (SEM, JEOL Ltd., Tokyo, Japan). The compositions were analyzed using an electron probe microanalyzer equipped with a wavelength-dispersive X-ray spectrometer (EPMA-WDS, JEOL Ltd., Tokyo, Japan). The crystal structures were determined by powder X-ray diffraction (XRD, Bruker Corp., Billerica, MA, USA) with Co-Kα radiation. For the XRD measurements, some of the bulk samples were crushed into powders, and were then sealed into quartz tubes. Strain-relief heat treatments at 1473 K for 2 min were performed on these powders, followed by quenching in ice water without breaking the tubes. Thermoanalysis and thermomagnetization measurements were used to determine the martensitic transformation and magnetic transition temperatures. The thermoanalysis was conducted using a DSC. The thermomagnetization curves were measured using a vibrating sample magnetometer (VSM, Toei Industry Co., Ltd., Tokyo, Japan), a superconducting quantum interference device (SQUID) magnetometer (Quantum Design Inc., San Diego, CA, USA), and the ac magnetic susceptibility (ACMS) option of the physical properties measurement system (PPMS, Quantum Design Inc., San Diego, CA, USA). Magnetization measurements up to 67.5 kOe were performed at 6 K by SQUID.

Magnetization measurements in the pulsed high magnetic fields, up to approximately 550 kOe, were performed at the Institute for Solid State Physics, the University of Tokyo. The pulsed magnetic fields have a duration of approximately 36 ms. To generate the maximum fields of this study, 550 kOe, a pulse magnet is driven by three condenser banks having total capacitance of 18 mF with the 9 kV charged voltage. The measurements were performed by the induction method using coaxial pickup coils. Refer to Reference [21] for further details.

3. Results and Discussion

3.1. Microstructure Observation

Figure 1 shows typical SEM micrographs of $Co_{64}V_{15}(Si_{21-x}Al_x)$ (xAl for short), captured by backscattered electrons (BSE), of the solution-heat-treated samples at room temperature. Figure 1a presents the microstructure of the martensite phase of the 0Al alloy with the residual parent phase. According to Jiang et al. [13], transmission electron microscope (TEM) observations of the $Co_{63.5}V_{17}Si_{19.5}$ alloy revealed a martensitic microstructure with a residual parent phase. Therefore, this result is consistent with that in the aforementioned paper. However, for 5Al and 12Al, as shown in Figure 1b,c, respectively, an almost full martensite single-phase microstructure was obtained. For the 13Al to 16Al alloys, the microstructure changed to a parent single-phase, as shown in Figure 1d. As shown in Figure 1e, among the prepared alloys, only the 21Al alloy shows a two-phase microstructure consisting of the matrix parent and a small amount of Co-rich precipitates ($Co_{77.2}V_{15.4}Al_{7.4}$). Each phase was identified by EPMA analysis and XRD measurements in Section 3.2.

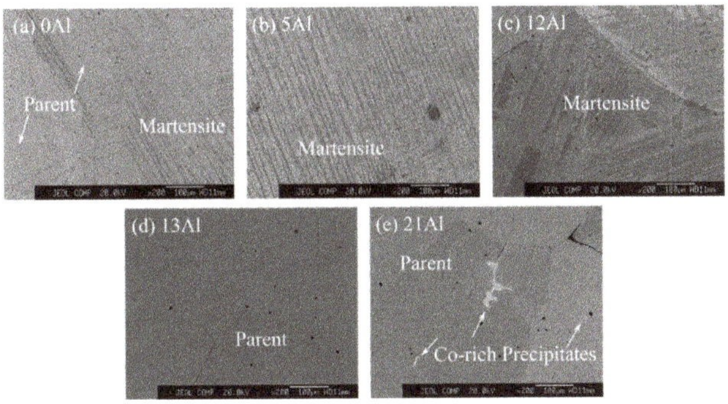

Figure 1. Micrographs taken by BSE of the (**a**) 0Al, (**b**) 5Al, (**c**) 12Al, (**d**) 13Al, and (**e**) 21Al alloys at room temperature. For (**a**) 0Al, parent and partial martensite phases were observed; (**b**) 5Al and (**c**) 12 Al exhibit a single martensite phase; and (**d**) 13Al has a single parent phase, whereas (**e**) 21Al contains Co-rich precipitates.

The compositions of the prepared alloys are summarized in Table 1. The Co and V concentrations in the alloys were found to be almost constant, and it is reasonable to plot the transformation temperatures within a pseudo-binary magnetic phase diagram.

Table 1. The chemical compositions of $Co_{64}V_{15}(Si_{21-x}Al_x)$ (xAl for short) alloys analyzed by an EPMA, where "P" and "M" mean parent phase and martensite phase, respectively.

Alloy	Phase at Room Temperature	Composition (Atomic %)			
		Co	V	Si	Al
0Al	P + M	64.0	15.4	20.6	-
5Al	M	63.6	15.4	16.5	4.5
9Al	M	64.2	15.5	11.3	9.0
10Al	M	64.1	15.5	10.1	10.2
11Al	M	64.3	15.7	9.1	10.9
12Al	M	63.8	15.6	8.3	12.3
13Al	P	64.2	15.7	7.1	13.0
14Al	P	64.2	15.7	6.1	13.9
15Al	P	64.2	15.5	5.3	15.0
16Al	P	64.8	15.7	3.7	15.8
21Al	P (Matrix)	64.5	15.5	-	20.0
	2nd	77.2	15.4	-	7.4

3.2. *Crystal Structures*

Figure 2a depicts the powder XRD patterns at room temperature. For 0Al and 13A to 21Al alloys, the B2 or $L2_1$ structure peaks, which indicate the parent phase [10], wer observed. The $L1_0$ or $D0_{22}$ structure peaks, which indicate the martensite phase, wer detected for 0Al to 13Al, as shown in Figure 2a. For the 21Al alloy, the peaks of the Co-ric precipitates were also found, as shown in Figure 2a. These peaks are consistent wit the SEM observations. For the alloys with a small Al concentration, TEM observation confirmed that the martensite phase had a $D0_{22}$ structure [19]. For the alloys with a larg Al concentration, the 111 and 200 superlattice reflection peaks for the $L2_1$ structure wer observed for the 13Al to 15Al alloys, as shown in Figure 2a. Furthermore, Figure 2 depicts an additional XRD scan focusing on the area between 40° and 60° for the 12Al allo whose martensitic transformation temperatures are near room temperature (T_{Ms} = 330 K i Section 3.3). No peak indicating a long-period stacking-ordered structure of martensite wa detected. Consequently, the crystal structure change in the martensitic transformation wa identified as the $L2_1/D0_{22}$ type for the $Co_{64}V_{15}Si_{21-x}Al_x$ system, which is the same as tha for other Co-based Heusler alloys, such as Co-Cr-Ga-Si, Co-Cr-Al-Si, and Co-V-Ga [10–12

In Figure 2a, some peaks originating from the residual parent phase or Co-rich pre cipitates appear for the 5Al to 12Al powder samples, which are not in agreement wit the microstructural observation shown in Figure 1. This may originate from the fact tha after the strain-relief heat treatments, the quartz tubes filled with powders were not broke when quenched in ice water, the cooling speed of which is slower than that of the bul samples. Consequently, the residual parent phase, as well as the Co-rich precipitate partially appeared after the strain-relief heat treatments. A similar phenomenon has bee reported for Co-Cr-Ga-Si [10] and Co-Cr-Al-Si alloys [11]. XRD measurements of bul samples were also conducted for the 0Al, 5Al, and 9Al alloys, of which the 5Al and 9A alloys were measured to compare with the powder XRD, and the results are shown i Figure 2c. Some peaks were missing, and the peak intensities were different from th calculated patterns, due to a small number of grains and/or the preferred textures in th bulk samples. However, the peaks of the $D0_{22}$ structure were found. For the 5Al and 9A the lattice constants are in good agreement with those determined by the powder and bul XRD measurements.

The lattice constants at room temperature are listed in Table 2. Figure 2d shows the composition dependence of the lattice constants for both the parent and martensite phases The lattice constants of the 0Al alloy were plotted from the bulk XRD patterns and the others were plotted from the powder XRD patterns. The $c/2a$ ratio, which indicates the tetragonality of the martensite, is also shown in the inset of Figure 2d. The lattice constant of the $L2_1$ phase increases with an increasing Al content, which may be attributed to the difference in the Si and Al atomic radii. The $c/2a$ ratio decreases when increasing the Al content. The composition dependence of the molar volume, which was calculated from the lattice constants determined at room temperature for each alloy, is shown in Figure 2e The value of the volume change between the parent and martensite phases was estimated to be up to 0.68% for the 12Al alloy (indicated by ΔV). For 0Al, the molar volume of the martensite phase was larger than that of the parent phase, which is a similar result to that of the $Co_{63.5}V_{17}Si_{19.5}$ alloy [13]. Note that these lattice parameters were determined only at room temperature; therefore, Figure 2e does not necessarily mean the volume change during the martensitic transformation, especially when the transformation temperature is high, such as in the case of the 0Al alloy.

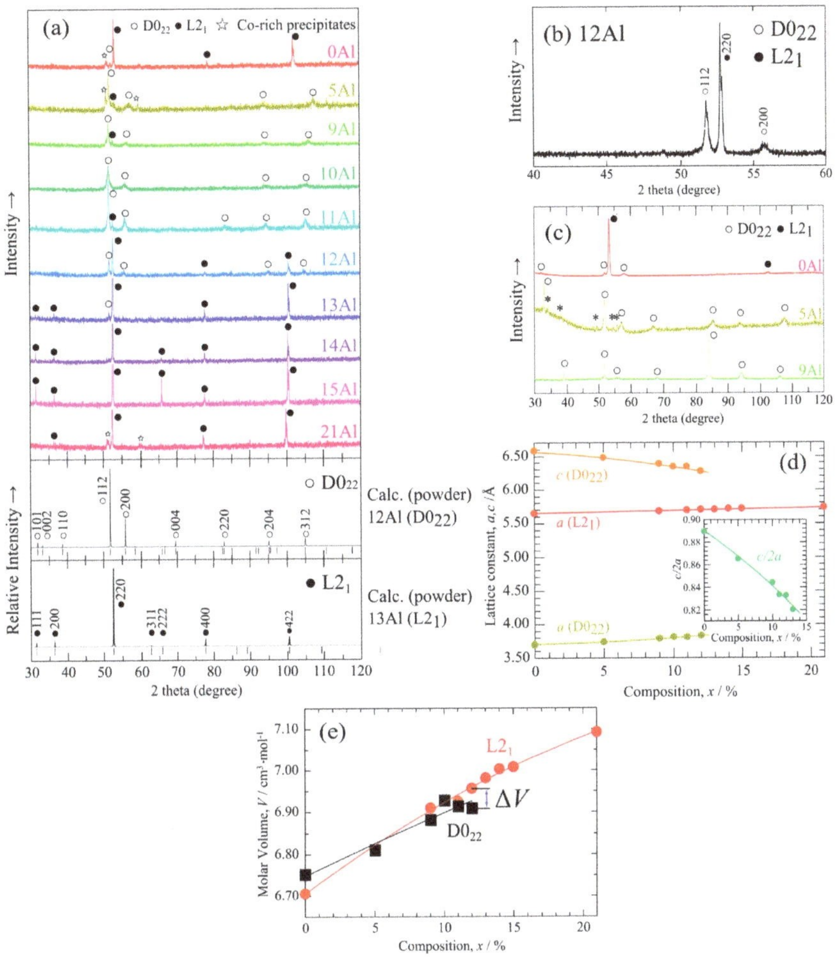

Figure 2. (**a**) Powder XRD patterns at room temperature for the $Co_{64}V_{15}Si_{21-x}Al_x$ system, and the calculated patterns for $Co_{64}V_{15}Si_9Al_{12}$ (D0$_{22}$) and $Co_{64}V_{15}Si_8Al_{13}$ (L2$_1$). (**b**) A detailed scan for the region near the main peaks of the 12Al alloy. (**c**) Bulk XRD patterns at room temperature for the 0Al, 5Al, and 9Al alloys. For the 5Al alloy, unknown peaks are marked with asterisks (*). (**d**) Composition dependence of the lattice constants for both the parent and martensite phases, together with (inset) the $c/2a$ ratio indicating the tetragonality of the martensite. (**e**) The composition dependence of the molar volume for both the parent and martensite phases. Solid lines in (**d**,**e**) are guides for the eye.

3.3. Determination of Martensitic and Magnetic Transformation Temperatures

Figure 3 shows the thermoanalysis results of the (a) 0Al, (b) 5Al and 9Al, and (c) 10Al to 13Al alloys in the $Co_{64}V_{15}Si_{21-x}Al_x$ system, where the peaks in the heating and cooling processes correspond to the martensitic transformation. Here, the forward martensitic transformation starting temperature (T_{Ms}) and the reverse martensitic transformation finishing temperature (T_{Af}) were evaluated using the extrapolation method, as shown in Figure 3. Table 3 summarizes the determined temperatures, where the thermal hysteresis of the martensitic transformation is defined as $T_{Af} - T_{Ms}$. The transformation temperatures decrease with an increasing Al content, and the T_{Ms} becomes lower than room temperature at 13Al, which is consistent with the electron microscopy observation shown in Figure 1. One may notice that the thermal hysteresis for the 10Al to 13Al alloys is less than 25 K

(Figure 3c), which is a typical feature of the thermoelastic martensitic transformation. In contrast, for the 0Al to 9Al alloys, the hysteresis is large, as shown in Figure 3a,b. As indicated by the dashed frames, some exothermic peaks at approximately 500 to 900 K, which may result from some diffusional transformation in the as-quenched specimen, were confirmed in the heating process. Thus, the martensite aging effect [22] is considered as a possible cause for the large hysteresis. As shown in Figure 3c, some alloys such as 13Al were found to show multiple peaks in the DSC curves, and similar phenomena have been reported in Ni-Mn-Al [23], Ni-Mn-In [24], and Ni-Co-Mn-Ga [25] alloys. This is a phenomenon well-known for alloys showing low martensitic transformation temperatures where the transformation entropy is small and the transformation interval ($T_{Ms} - T_{Mf}$) is large. However, we still cannot rule out the possibility of an intermartensitic transformation as in Ni-Mn-Ga alloy [26]. Thus, a follow-up study may be required to clarify this problem.

Table 2. Lattice constants determined for the parent and martensite phases at room temperature. A hyphen indicates that the phases do not exist, or that the lattice constants could not be determined. Unmeasured data are left blank.

Alloy	Lattice Constant (Powder)				Lattice Constant (Bulk)			
	a ($L2_1$, nm)	*a* ($D0_{22}$, nm)	*c* ($D0_{22}$, nm)	*c*/2*a*	*a* ($L2_1$, nm)	*a* ($D0_{22}$, nm)	*c* ($D0_{22}$, nm)	*c*/2*a*
0Al	0.5627	-	-	-	0.5628	0.3693	0.6579	0.8909
5Al	-	0.3739	0.6472	0.8654	-	0.3739	0.6472	0.8654
9Al	0.5684	0.3783	0.6389	0.8444	-	0.3781	0.6399	0.8462
10Al	-	0.3808	0.6348	0.8335	-	-	-	-
11Al	0.5688	0.3806	0.6341	0.8330	-	-	-	-
12Al	0.5696	0.3825	0.6275	0.8203	-	-	-	-
13Al	0.5703	-	-	-	-	-	-	-
14Al	0.5709	-	-	-	-	-	-	-
15Al	0.5710	-	-	-	-	-	-	-
21Al	0.5733	-	-	-	-	-	-	-

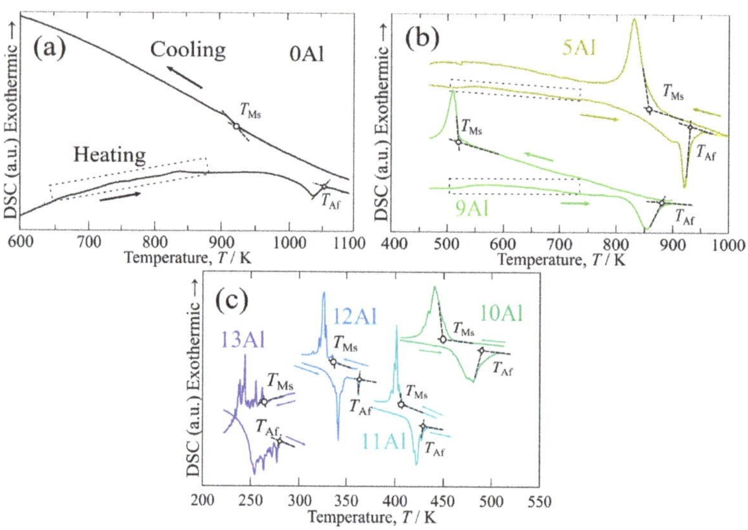

Figure 3. Thermoanalysis by DSC for the (**a**) 0Al, (**b**) 5Al and 9Al, and (**c**) 10Al to 13Al alloys.

Table 3. Summary of magnetic and martensitic transformation temperatures determined by DSC, VSM, ACMS, and SQUID. $T_C{}^P$ is the Curie temperature of the parent phase and $T_C{}^M$ is that of martensite phase. T_{Ms} and T_{Af} are the martensitic transformation starting temperature and the reverse martensitic transformation finishing temperature, respectively. The thermal hysteresis of the martensitic transformation is defined as $T_{Af} - T_{Ms}$. A hyphen indicates that the transformation does not occur or cannot be determined. Unmeasured data are left blank.

Alloy	Transformation Temperatures (K)				Thermal Hysteresis
	$T_C{}^P$	$T_C{}^M$	T_{Ms}	T_{Af}	$T_{Af} - T_{Ms}$ (K)
0Al	-		928	1056	128
5Al	-	27	836	923	88
9Al	-	92	518	882	364
10Al	-	-	449	491	24
11Al	-	-	405	430	25
12Al	-	-	330	348	18
13Al	-	-	265	280	15
14Al	-	160	185	209	24
15Al	195	-	90	108	18
16Al	202	-	-	-	-
21Al	210	-	-	-	-

Figure 4 shows the thermomagnetization curves of the xAl alloys under a magnetic field of 500 Oe. For the 14Al and 15Al alloys, the thermal hysteresis associated with the martensitic transformation was confirmed, where T_{Ms} and T_{Af} were determined by extrapolation. For the 15Al to 21Al alloys, the Curie temperature of the parent phase ($T_C{}^P$) was observed to slightly increase with an increasing Al composition. Moreover, the Curie temperature of the martensite phase ($T_C{}^M$) was observed to be below the T_{Ms} for the 5Al, 9Al, and 14Al alloys, which was also observed in the Ni-Mn-X (X = In, Sn, Sb [7], and Ga [27]) Heusler alloys, and the Co_2NbSn Heusler alloy [9]. The Curie temperature of the martensite phase increases with increasing the Al content. Table 3 summarizes the transformation temperatures for the thermomagnetization measurements.

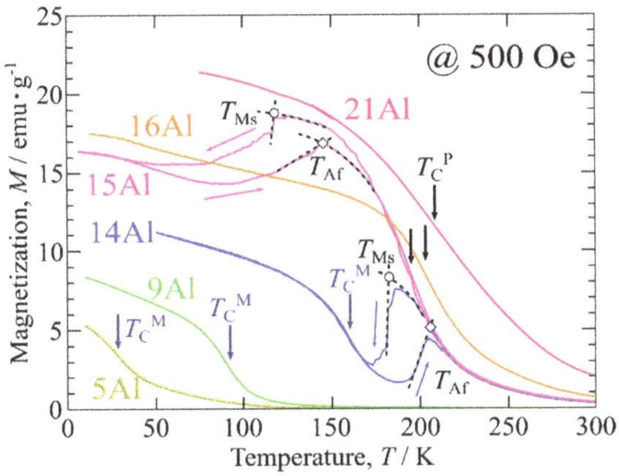

Figure 4. Thermomagnetization curves for the $Co_{64}V_{15}Si_{21-x}Al_x$ (xAl) alloys under a magnetic field of 500 Oe. The martensitic transformation temperatures ($\boldsymbol{T_{Ms}}$, $\boldsymbol{T_{Af}}$) were determined for the 14Al and 15Al alloys. Additionally, the Curie temperatures of the martensite ($T_C{}^M$) and parent ($T_C{}^P$) phases were observed for the 5Al to 14Al and 15Al to 21Al alloys, respectively.

The data listed in Table 3 are plotted as a pseudo-binary magnetic phase diagram in Figure 5 and the $Co_{64}V_{15}Si_{17}Al_4$ data reported by Zhang et al. [19] are also shown for comparison. The T_{Ms} and T_{Af} decrease with increasing Al from over 900 K (627 °C) to below room temperature. The magnetic phase diagram determined for the $Co_{64}V_{15}Si_{21-x}Al_x$ pseudo-binary system is similar to that for the $Co_xV_{(100-x)/2}Ga_{(100-x)/2}$ [12] and $Ni_{50}Mn_{50-x}Ga_x$ [27] systems because the martensitic transformation temperature can be extensively changed.

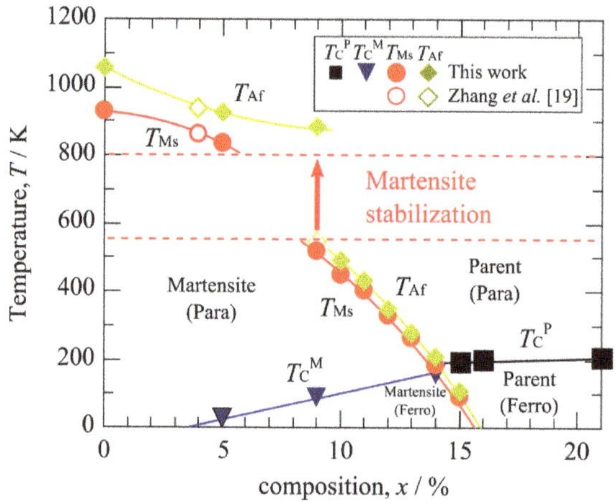

Figure 5. Pseudo-binary magnetic phase diagram of the $Co_{64}V_{15}Si_{21-x}Al_x$ system, where "Para" and "Ferro" mean paramagnetic and ferromagnetic, respectively. The martensitic transformation temperatures reported by Zhang et al. [19] are plotted by open circle and open rhombus and the solid lines are guides for the eye.

3.4. Magnetic Properties

Figure 6a depicts the magnetization curves at 6 K for the 5Al to 21Al alloys. At 6 K, the alloys from 5Al to 15Al were in the martensite phase, whereas the 16Al and 21Al alloys in the parent phase, as shown in the magnetic phase diagram in Figure 5. The values of spontaneous magnetization were determined using the Arrott plot method [28] and they are summarized in Table 4 and plotted against the Al composition in Figure 6b. When the Al composition increases, both the spontaneous magnetization and Curie temperature increase linearly, as shown in Figure 6b. Furthermore, as indicated by ΔM and ΔT_C discontinuous jumps in both the spontaneous magnetization and the Curie temperature can be seen around the 15Al alloy, which reflects the difference in magnetization of the parent and martensite phases. The difference in the magnetization is more pronounced than that of the $Ni_{50}Mn_{50-x}Ga_x$ alloys [27]; however, it is less obvious than that of the $Ni_{50}Mn_{50-x}In_x$ alloys [29].

Table 4. Spontaneous magnetization at 6 K.

Alloy	Spontaneous Magnetization	
	emu g^{-1}	μ_B f.u.$^{-1}$
5Al	10.8	0.39
9Al	18.6	0.68
14Al	28.8	1.05
15Al	30.5	1.11
16Al	35.0	1.28
21Al	40.4	1.48

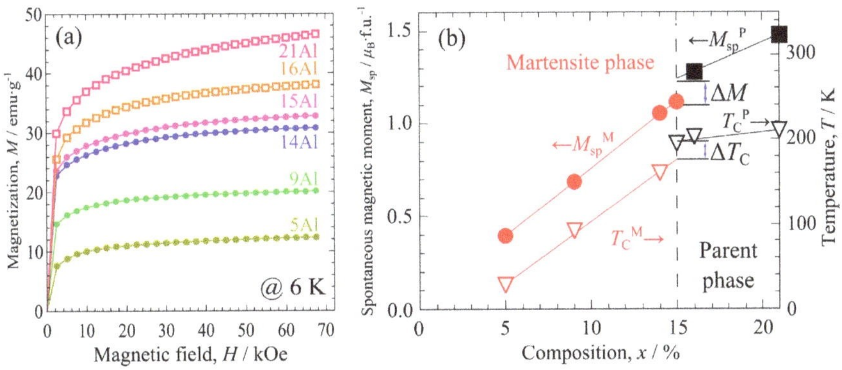

Figure 6. (**a**) Magnetization curves at 6 K for the 5Al to 21Al alloys. At 6 K, the alloys from 5Al to 15Al are in the martensite phase (plotted by filled circles), whereas the 16Al and 21Al alloys are in the parent phase (plotted by the open squares). (**b**) The spontaneous magnetization evaluated from the magnetization curves in (**a**) (filled marks). The Curie temperatures of the parent (T_C^P) and martensite (T_C^M) phases are also plotted (open marks). Solid lines are guides for the eye.

3.5. Magnetic-Field-Induced Reverse Martensitic Transformation

For the 14Al alloy, a change in magnetization due to the martensitic transformation was observed during the thermomagnetization measurement, as shown in Figure 4. Thus, one can expect a magnetic-field-induced reverse martensitic transformation. Here, magnetization measurements, using pulsed high magnetic fields up to approximately 550 kOe, were performed. Figure 7a depicts the results. A magnetic-field-induced reverse martensitic transformation was observed at all measured temperatures. Except for 140 K, almost the full parent phase was induced before the magnetic field reached 550 kOe; when the magnetic field decreased, the parent phase transformed back to the martensite phase.

The critical magnetic fields, the forward martensitic transformation starting magnetic field H_{Ms}, and the reverse martensitic transformation finishing magnetic field H_{Af}, were determined by the extrapolation method, as depicted in Figure 7b. The H_{Ms} and H_{Af} are plotted against the measurement temperature in Figure 7c, where $H_0 = (H_{Ms} + H_{Af})/2$ was assumed to be the thermodynamic equilibrium magnetic field. For comparison, the results of the Co-Cr-Ga-Si and Ni-Co-Mn-In alloys are also plotted [15,30]. It was confirmed that the critical magnetic fields decreased almost linearly with an increasing measurement temperature. The linear fits in Figure 7c were used to estimate the temperature dependence of the equilibrium magnetic field dH_0/dT, and the entropy change ΔS could be calculated using the Clausius–Clapeyron equation:

$$\frac{dH_0}{dT} = -\frac{\Delta S}{\Delta M}, \tag{1}$$

where ΔS and ΔM are the entropy change and the magnetization difference during the martensitic transformation, respectively. The magnetization difference, ΔM, was assumed as 12.8 emu·g^{-1} for simplicity, as shown in Figure 7a,b. Therefore, the entropy change, ΔS, was calculated and plotted in Figure 7d. Additionally, the entropy change was also evaluated from the DSC measurements in Figure 3c as:

$$\Delta S = \frac{\Delta H_{M \to P}}{T_p}, \tag{2}$$

where ΔH is the latent heat and T_p is the peak temperature in the reverse martensitic transformation. Figure 7d shows that the estimated entropy changes, $-\Delta S$, from the magnetization and DSC measurements, shows a same temperature dependence. For comparison, the $-\Delta S$ of Ti-Ni, Cu-Al-Mn, Fe-Mn-Al-Cr-Ni, and Co-Cr-Ga-Si alloys [15,31,32] are also plotted in Figure 7d. The $-\Delta S$ of the current Co-V-Si-Al alloys is significantly smaller than those of the Ti-Ni and Cu-Al-Mn alloys in the temperature range of 160 to 500 K. Using the small transformation entropy at a wide temperature range, superelasticity, like in the Fe-Mn-Al-Ni-based [32,33] alloys which show a small temperature dependence of critical stress and operate in a wide temperature range, may be realized with the Co-V-Si-Al alloys.

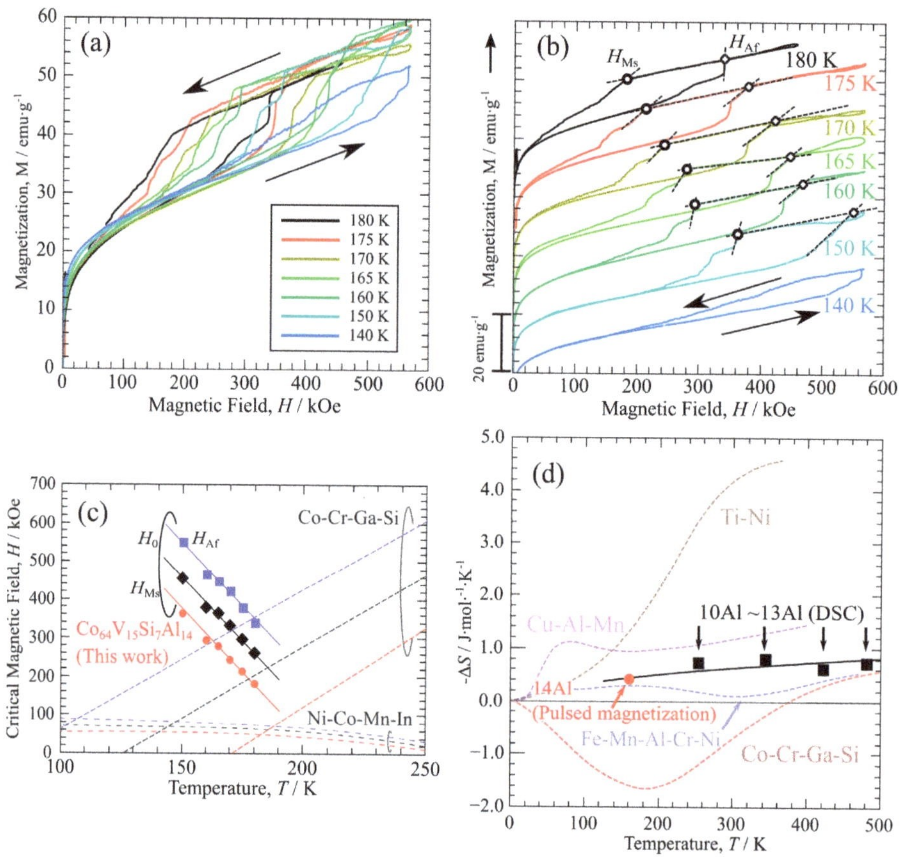

Figure 7. (**a**,**b**) Magnetization curves of the $Co_{64}V_{15}Si_7Al_{14}$ (14Al) alloy under pulsed magnetic fields up to approximately 550 kOe. A magnetic-field-induced reverse martensitic transformation was observed. The forward martensitic transformation starting magnetic field (H_{Ms}) and the reverse martensitic transformation finishing magnetic field (H_{Af}) were defined by the extrapolation method, as shown in (**b**,**c**) H_{Ms}, H_{Af}, and $H_0 = (H_{Ms} + H_{Af})/2$ are plotted against the measurement temperature. The solid lines are the least-square fits. The critical magnetic fields of the Co-Cr-Ga-Si [15] and Ni-Co-Mn-In [30] alloys are also shown by dashed lines. (**d**) Entropy changes, ΔS, were estimated using the Clausius-Clapeyron equation for the 14Al alloy and by thermoanalysis for the 10Al to 13Al alloys, together with that of the Ti-Ni, Cu-Al-Mn [31], Fe-Mn-Al-Cr-Ni [32], and Co-Cr-Ga-Si alloys [15] shown by dashed lines.

4. Conclusions

In this work, the crystal structures, martensitic transformation behavior, magnetic properties, and metamagnetic transition were investigated for $Co_{64}V_{15}(Si_{21-x}Al_x)$ alloys. The XRD measurement results revealed that the crystal structures of the martensite and the parent phases are $D0_{22}$ and $L2_1$, respectively, which are the same as those of the other Co-based Heusler alloys showing martensitic transformations.

A pseudo-binary magnetic phase diagram of the $Co_{64}V_{15}Si_{21-x}Al_x$ system was determined. The martensitic transformation temperature decreases with increasing Al compositions from over 900 K to below room temperature. Both the Curie temperature of the parent and martensite phases were observed. The composition dependences of the spontaneous magnetization and the Curie temperature were investigated, and discontinuous jumps in both the spontaneous magnetization and the Curie temperature were observed, which reflects the difference in magnetization of the parent and martensite phases.

The magnetic-field-induced reverse martensitic transformation of the $Co_{64}V_{15}Si_7Al_{14}$ alloy was observed by magnetization measurements by using pulsed high magnetic fields up to 550 kOe. Moreover, the temperature dependence of the transformation entropy changes of the Co-V-Si-Al alloys was estimated from the results of the DSC and the pulsed magnetization measurements. Compared to the Ti-Ni and Cu-Al-Mn shape memory alloys, the transformation entropy of the Co-V-Si-Al alloys is significantly smaller in the temperature range of 160 and 500 K.

Author Contributions: Conceptualization, X.X., T.O., and R.K.; investigation, K.N. and A.M.; resources, R.K. and M.T.; writing—original draft preparation, K.N.; writing—review and editing, A.M., X.X., T.O., M.T., and R.K. All authors have read and agreed to the published version of the manuscript.

Funding: This study was partially supported by Grants-in-Aid (No. 19H02412 and No. 15H05766) from the Japan Society for the Promotion of Science (JSPS).

Data Availability Statement: The data presented in this study are available on request from the corresponding author.

Acknowledgments: A part of the experiments were performed at the Center for Low Temperature Science, Institute for Materials Research, Tohoku University.

Conflicts of Interest: The authors declare no conflict of interest.

References

1. Buehler, W.J.; Gilfrich, J.V.; Wiley, R.C. Effect of low-temperature phase changes on the mechanical properties of alloys near composition TiNi. *J. Appl. Phys.* **1963**, *34*, 1475–1477. [CrossRef]
2. Kainuma, R.; Takahashi, S.; Ishida, K. Thermoelastic martensite and shape memory effect in ductile Cu-Al-Mn alloys. *Metall. Mater. Trans. A Phys. Metall. Mater. Sci.* **1996**, *27*, 2187–2195. [CrossRef]
3. Omori, T.; Ando, K.; Okano, M.; Xu, X.; Tanaka, Y.; Ohnuma, I.; Kainuma, R.; Ishida, K. Superelastic Effect in Polycrystalline Ferrous Alloys. *Science* **2011**, *333*, 68–71. [CrossRef] [PubMed]
4. Ogawa, Y.; Ando, D.; Sutou, Y.; Koike, J. A lightweight shape-memory magnesium alloy. *Science* **2016**, *353*, 368–370. [CrossRef]
5. Ullakko, K.; Huang, J.K.; Kantner, C.; O'Handley, R.C.; Kokorin, V.V. Large magnetic-field-induced strains in Ni_2MnGa single crystals. *Appl. Phys. Lett.* **1996**, *69*, 1966–1968. [CrossRef]
6. Kainuma, R.; Imano, Y.; Ito, W.; Sutou, Y.; Morito, H.; Okamoto, S.; Kitakami, O.; Oikawa, K.; Fujita, A.; Kanomata, T.; et al. Magnetic-field-induced shape recovery by reverse phase transformation. *Nature* **2006**, *439*, 957–960. [CrossRef]
7. Sutou, Y.; Imano, Y.; Koeda, N.; Omori, T.; Kainuma, R.; Ishida, K.; Oikawa, K. Magnetic and martensitic transformations of NiMnX(X=In, Sn, Sb) ferromagnetic shape memory alloys. *Appl. Phys. Lett.* **2004**, *85*, 4358–4360. [CrossRef]
8. Galanakis, I.; Dederichs, P.H.; Papanikolaou, N. Slater-Pauling behavior and origin of the half-metallicity of the full-Heusler alloys. *Phys. Rev. B* **2002**, *66*, 174429. [CrossRef]
9. Terada, M.; Fujita, Y.; Endo, K. Magnetic properties of the Heusler alloys M_2XSn (M=Co or Ni, X=Zr, Nb or Hf). *J. Phys. Soc. Jpn.* **1974**, *36*, 620. [CrossRef]
10. Xu, X.; Omori, T.; Nagasako, M.; Okubo, A.; Umetsu, R.Y.; Kanomata, T.; Ishida, K.; Kainuma, R. Cooling-induced shape memory effect and inverse temperature dependence of superelastic stress in Co_2Cr(Ga,Si) ferromagnetic Heusler alloys. *Appl. Phys. Lett.* **2013**, *103*, 164104. [CrossRef]
11. Hirata, K.; Xu, X.; Omori, T.; Nagasako, M.; Kainuma, R. Martensitic transformation and superelasticity in off-stoichiometric Co_2Cr(AlSi) Heusler alloys. *J. Alloys Compd.* **2015**, *642*, 200–203. [CrossRef]
12. Xu, X.; Nagashima, A.; Nagasako, M.; Omori, T.; Kanomata, T.; Kainuma, R. Martensitic transformation and phase diagram in ternary Co-V-Ga Heusler alloys. *Appl. Phys. Lett.* **2017**, *110*, 121906. [CrossRef]
13. Jiang, H.; Xu, X.; Omori, T.; Nagasako, M.; Ruan, J.; Yang, S.; Wang, C.; Liu, X.; Kainuma, R. Martensitic transformation and shape memory effect at high temperatures in off-stoichiometric Co_2VSi Heusler alloys. *Mater. Sci. Eng. A* **2016**, *676*, 191–196. [CrossRef]

14. Jiang, H.; Yang, S.; Wang, C.; Zhang, Y.; Xu, X.; Chen, Y.; Omori, T.; Kainuma, R.; Liu, X. Martensitic transformation and shap memory effects in Co-V-Al alloys at high temperatures. *J. Alloys Compd.* **2019**, *786*, 648–654. [CrossRef]
15. Xu, X.; Kihara, T.; Miyake, A.; Tokunaga, M.; Kanomata, T.; Kainuma, R. Magnetic-field-induced transition for reentrar martensitic transformation in Co-Cr-Ga-Si shape memory alloys. *J. Magn. Magn. Mater.* **2018**, *466*, 273–276. [CrossRef]
16. Odaira, T.; Xu, X.; Miyake, A.; Omori, T.; Tokunaga, M.; Kainuma, R. Thermal, magnetic field- and stress-induced transformatio in Heusler-type Co-Cr-Al-Si shape memory alloys. *Scripta Mater.* **2018**, *153*, 35–39. [CrossRef]
17. Liu, C.; Li, Z.; Zhang, Y.; Huang, Y.; Ye, M.; Sun, X.; Zhang, G.; Cao, Y.; Xu, K.; Jing, C. Realization of metamagnetic martensiti transformation with multifunctional properties in $Co_{50}V_{34}Ga_{16}$ Heusler alloy. *Appl. Phys. Lett.* **2018**, *112*, 211903. [CrossRef]
18. Jiang, H.; Wang, C.; Xu, W.; Xu, X.; Yang, S.; Kainuma, R.; Liu, X. Alloying effects of Ga on the Co-V-Si high-temperature shap memory alloys. *Mater. Des.* **2017**, *116*, 300–308. [CrossRef]
19. Zhang, Y.; Jiang, H.; Liu, X.; Huang, L.; Yang, S.; Wang, C. Thermal cycle stability of $Co_{64}V_{15}Si_{17}Al_4$ high-temperature shap memory alloy. *Mater. Lett.* **2020**, *260*, 126930. [CrossRef]
20. Odaira, T.; Xu, S.; Xu, X.; Omori, T.; Kainuma, R. Elastocaloric switching effect induced by reentrant martensitic transformatio *Appl. Phys. Rev.* **2020**, *7*, 031406. [CrossRef]
21. Kindo, K. 100T magnet developed in Osaka. *Phys. B Condens. Matter* **2001**, *294–295*, 585–590. [CrossRef]
22. Otsuka, K.; Ren, X. Mechanism of martensite aging effect. *Scripta Mater.* **2004**, *50*, 207–212. [CrossRef]
23. Kainuma, R.; Ishida, K.; Nakano, H. Martensitic transformations in NiMnAl β phase alloys. *Metall. Mater. Trans. A* **1996**, *2* 4153–4162. [CrossRef]
24. Krenke, T.; Duman, E.; Acet, M.; Wassermann, E.F.; Moya, X.; Mañosa, L.; Planes, A.; Suard, E.; Ouladdiaf, B. Magneti superelasticity and inverse magnetocaloric effect in Ni-Mn-In. *Phys. Rev. B* **2007**, *75*, 104414. [CrossRef]
25. Xu, X.; Ito, W.; Umetsu, R.Y.; Koyama, K.; Kainuma, R.; Ishida, K. Kinetic arrest of martensitic transformation i $Ni_{33.0}Co_{13.4}Mn_{39.7}Ga_{13.9}$ metamagnetic shape memory alloy. *Mater. Trans.* **2010**, *51*, 469–471. [CrossRef]
26. Sozinov, A.; Likhachev, A.A.; Lanska, N.; Ullakko, K. Giant magnetic-field-induced strain in NiMnGa seven-layered martensiti phase. *Appl. Phys. Lett.* **2002**, *80*, 1746–1748. [CrossRef]
27. Xu, X.; Nagasako, M.; Ito, W.; Umetsu, R.Y.; Kanomata, T.; Kainuma, R. Magnetic properties and phase diagram of $Ni_{50}Mn_{50-x}Ga_x$ ferromagnetic shape memory alloys. *Acta Mater.* **2013**, *61*, 6712–6723. [CrossRef]
28. Arrott, A. Criterion for ferromagnetism from observations of magnetic isotherms. *Phys. Rev.* **1957**, *108*, 1394–1396. [CrossRef]
29. Kanomata, T.; Yasuda, T.; Sasaki, S.; Nishihara, H.; Kainuma, R.; Ito, W.; Oikawa, K.; Ishida, K.; Neumann, K.U.; Ziebeck, K.R.A Magnetic properties on shape memory alloys $Ni_2Mn_{1+x}In_{1-x}$. *J. Magn. Magn. Mater.* **2009**, *321*, 773–776. [CrossRef]
30. Ito, W.; Ito, K.; Umetsu, R.Y.; Kainuma, R.; Koyama, K.; Watanabe, K.; Fujita, A.; Oikawa, K.; Ishida, K.; Kanomata, T. Kineti arrest of martensitic transformation in the NiCoMnIn metamagnetic shape memory alloy. *Appl. Phys. Lett.* **2008**, *92*, 02190 [CrossRef]
31. Niitsu, K.; Kimura, Y.; Omori, T.; Kainuma, R. Cryogenic superelasticity with large elastocaloric effect. *NPG Asia Mater.* **2018** *10*, e457. [CrossRef]
32. Xia, J.; Noguchi, Y.; Xu, X.; Odaira, T.; Kimura, Y.; Nagasako, M.; Omori, T.; Kainuma, R. Iron-based superelastic alloys with near-constant critical stress temperature dependence. *Science* **2020**, *369*, 855–858. [CrossRef] [PubMed]
33. Xia, J.; Xu, X.; Miyake, A.; Kimura, Y.; Omori, T.; Tokunaga, M.; Kainuma, R. Stress- and magnetic field-induced martensitic transformation at cryogenic temperatures in Fe–Mn–Al–Ni shape memory alloys. *Shape Mem. Superelasticity* **2017**, *3*, 467–475 [CrossRef]

metals

MDPI

Article

Influence of Structural Defects on the Properties of Metamagnetic Shape Memory Alloys

J. I. Pérez-Landazábal [1,2,*], V. Sánchez-Alarcos [1,2], V. Recarte [1,2], O. A. Lambri [3], F. G. Bonifacich [3], D. L.R. Khanna [1,2], I. Unzueta [4], J.A. García [5,6], F. Plazaola [7], J. López-García [1,8], M. Jimenez Ruiz [8], J.A. Rodríguez-Velamazán [8] and E. Cesari [9]

1 Departamento de Ciencias, Public University of Navarra (UPNa), 31006 Pamplona, Spain; vicente.sanchez@unavarra.es (V.S.-A.); recarte@unavarra.es (V.R.); deepali.khanna@unavarra.es (D.L.R.K.); javier.lopezg@unavarra.es (J.L.-G.)

2 Institute for Advanced Materials and Mathematics INAMAT^2, Universidad Pública de Navarra, 31006 Pamplona, Spain

3 Laboratorio de Materiales, Escuela de Ingeniería Eléctrica, Centro de Tecnología e Investigación Eléctrica, Facultad de Ciencias Exactas, Ingeniería y Agrimensura, Universidad Nacional de Rosario—CONICET, Avda. Pellegrini 250, 2000 Rosario, Argentina; olambri@fceia.unr.edu.ar (O.A.L.); bonifaci@fceia.unr.edu.ar (F.G.B.)

4 Matematika Aplikatua Saila, Bilboko Ingenieritza Eskola, Euskal Herriko Unibersitatea UPV/EHU, p.k. 644, 48080 Bilbao, Spain; iraultza.unzueta@ehu.eus

5 Fisika Aplikatua II Saila, Euskal Herriko Unibersitatea UPV/EHU, p.k. 644, 48080 Bilbao, Spain; joseangel.garcia@ehu.es

6 BC Materials (Basque Centre for Materials, Applications and Nanostructures), 48080 Leioa, Spain

7 Elektrizitate eta Elektronika Saila, Euskal Herriko Unibersitatea UPV/EHU, p.k. 644, 48080 Bilbao, Spain; fernando.plazaola@ehu.es

8 Institut Laue Langevin, 71, Avenue des Martyrs, 38042 Grenoble CEDEX, France; jimenez@ill.fr (M.J.R.); velamazan@ill.eu (J.A.R.-V.)

9 Departament de Física, Universitat de les Illes Balears, Ctra. de Valldemossa, km 7.5, E-07122 Palma de Mallorca, Spain; eduard.cesari@uib.cat

* Correspondence: ipzlanda@unavarra.es; Tel.: +34-948168448

Received: 17 July 2020; Accepted: 20 August 2020; Published: 22 August 2020

Abstract: The production of μ-particles of Metamagnetic Shape Memory Alloys by crushing and subsequent ball milling process has been analyzed. The high energy involved in the milling process induces large internal stresses and high density of defects with a strong influence on the martensitic transformation; the interphase creation and its movement during the martensitic transformation produces frictional contributions to the entropy change (exothermic process) both during forward and reverse transformation. The frictional contribution increases with the milling time as a consequence of the interaction between defects and interphases. The influence of the frictional terms on the magnetocaloric effect has been evidenced. Besides, the presence of antiphase boundaries linked to superdislocations helps to understand the spin-glass behavior at low temperatures in martensite. Finally, the particles in the deformed state were introduced in a photosensitive polymer. The mechanical damping associated to the Martensitic Transformation (MT) of the particles is clearly distinguished in the produced composite, which could be interesting for the development of magnetically-tunable mechanical dampers.

Keywords: metamagnetic shape memory alloys; structural defects; magnetocaloric effect; mechanical damping

1. Introduction

The change in the interatomic distances caused by the occurrence of a thermoelastic Martensitic Transformation (MT) in Metamagnetic Shape Memory Alloys (MSMA) results in large magnetization changes (ΔM) at the transformation temperature [1–4]. In fact, the strong dependence of the magnetic exchange interactions on the Mn-Mn distances [5–7] is responsible for such a change and allows the induction of the MT by external magnetic fields. The large magnetoresistance or the giant inverse magnetocaloric properties are among others, relevant effects for sensing and magnetic refrigeration [8–14] that make these alloys interesting from the application point of view. Besides, the inherent damping properties of shape memory alloys can be used for active noise reduction and mechanical damping applications [15,16]. The damping in the case of MSMA, can be controlled by an external magnetic field through the hysteretic induced MT. In particular, Ni-Mn-In and Ni-Mn-Sn show a MT between a ferromagnetic austenite and a weak-magnetic martensite. The compositional [17–20] and atomic order [21–24] influence on the MT properties has been widely studied in the literature. Although the long-range atomic order in Ni-Mn-In alloys can be easily controlled by means of thermal treatments, the Ni-Mn-Sn system shows a highly stable $L2_1$ structure where the atomic order can hardly be modified by conventional thermal treatments. The addition of Cobalt in both cases enhances the ferromagnetic ordering of the austenite and reduces the magnetism in martensite in Ni-Mn-X alloys, thus leading to an increase of ΔM and therefore to larger magnetically-induced shifts of the MT temperature [1,25–28].

On the other side, the new Additive Manufacturing (AM) technology (3D printing) opens a huge field in devices design, for example using polymers with different percentages of functional particles. In this context, the metallic particle production, their characterization and properties optimization are milestones to get valuable functional composites [29–33]. Mechanical milling is a simple method to produce micro and nanoparticles but unfortunately produces huge internal stress, defects and deformations that strongly modifies properties of the bulk alloy [34–39]. As an example, the high density of the anti-phase boundaries linked to dislocations produced during the milling process [40] or the local atomic disordering in boundary region [29] promotes antiferromagnetic coupling between Mn moments and consequently a reduction in the saturation magnetization of both martensitic and austenitic phases.

In this work, a hand crushing and subsequent ball milling process has been tested to produce $Ni_{45}Co_5Mn_{37}In_{13}$ μ-particles to be introduced in a photosensitive polymer. The high energy involved in the process produces large internal stresses and high density of defects with a strong influence on the MT. The work is focused on the analysis of the influence of milling on the structural and magnetic properties of MSMA. The frictional contribution to the MT entropy change has been shown to increase with the milling time. Besides, the presence of antiphase boundaries linked to superdislocations helps to understand a spin-glass behavior found at low temperatures. A prospective study for future applications has been analyzed; in particular, the refrigeration capacity of the μ-particles and the damping properties of the designed composites has been tested.

2. Materials and Methods

$Ni_{45}Co_5Mn_{36}In_{14}$ Metamagnetic alloys were synthetized from high purity elements by arc-melting under protective argon atmosphere. The bulks obtained were homogenized at 1170 K during 24 h. The composition of the alloys was analyzed by EDS in a JSM-5610LV Scanning Electron Microscope (SEM, Jeol, Tokio, Japan). The alloys were milled in an agate mortar until reaching a uniform particle-size distribution. Subsequently, a systematic ball milling was performed at Room Temperature (RT) and at different milling times (15, 30, 45, 60, 90 min) in a Retsch PM4 under argon atmosphere and controlled temperature, avoiding the possibility of oxidation. As-crushed powders were used to produce the MSMA μParticle-Polymers composites. The polymeric matrix was a commercial photo curable bisphenol A-glycidyl methacrylate (Bis-GMA) resin (Schmidt Composite flow Madrid, Spain). The MT was characterized by differential scanning calorimetry (Q-100 Differential Scanning Calorimeter,

DSC, TA Instruments, New Castle, DE, USA) at a heating/cooling rate of 10 K/min. The macroscopic magnetic characterization of the alloys was performed by SQUID magnetometry (QD MPMS XL-7, Quantum Design, San Diego, CA, USA) under different fields from 100 Oe to 60 kOe. High-Resolution X-rays Powder Diffraction (HRXPD,) was carried out in the BL04_MPSD beamline (28 keV) at ALBA synchrotron, Barcelona, Spain using a high-resolution detector (mad26) [41,42]. Diffraction data treatment was carried out using the programs of the Fullprof Suite [43]. Dynamic Mechanical Analysis (DMA) measurements, loss tangent (damping), tan(ϕ), and dynamic shear modulus, G′, were carried out as a function of temperature in a homemade torsion pendulum at frequencies close to 5 Hz. Measurements were performed during heating and cooling with a rate of 1 K/min. Measurements were performed under Argon atmosphere at atmospheric pressure.

3. Results and Discussion

Figure 1 shows the DSC thermograms performed on cooling/heating in the bulk, hand-crushed and ball-milled samples. The occurrence of a first-order MT is evidenced by the presence of the hysteretic exothermic (forward MT) and endothermic (reverse MT) peaks at around 275 K and 295 K respectively. A second-order magnetic transition taking place in the austenitic phase can be inferred from the λ-type shoulder linked to the Curie temperature observed around T_c = 388 K. Focusing on the MT peak, the area below the peak clearly decreases with deformation. Nevertheless, the peak position does not change, indicating that the long-range atomic order does not evolve with deformation, in agreement with the inalterability of the Curie temperature. Figure 1 right shows the evolution of the enthalpy change at the MT with the milling time. A clear decrease is observed for the hand-crushed sample, and then a gradual decrease of the enthalpy takes place as the milling time increases (after 60 min the MT practically disappears). The easiest way to understand this behavior is assuming a reduction in the transforming volume fraction as a consequence of the induced internal stresses (defects, deformations) during hand crushing or milling processes. Nevertheless, deformation itself can also contribute to the transformation enthalpies and entropies.

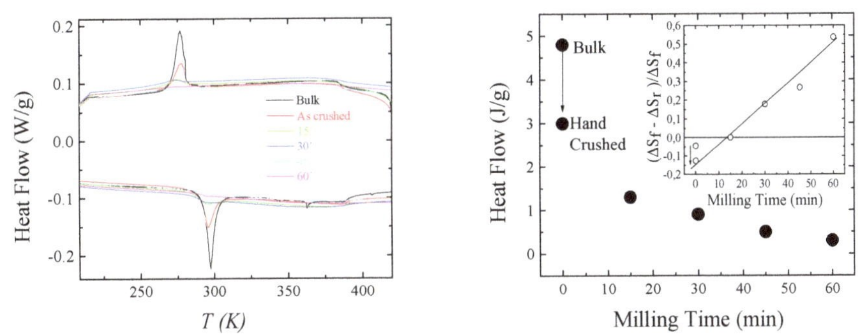

Figure 1. DSC measurements around the MT for bulk, as-crushed, and milled sample (**left**). The evolution of the heat flow (forward MT) as a function of the milling time is shown on the **right** side (the inset shows the normalized difference between forward and reverse entropies).

The entropy change linked to the MT, which limits the magnetocaloric effect, can be considered as the sum of a vibrational, ΔS_{vib}, and a magnetic, ΔS_{mag}, terms (the small electronic term is usually neglected), in such a way that $\Delta S^{O} \approx \Delta S_{vib} + \Delta S_{mag}$ [44] (the superscript refers to the ideal state, where the frictional contribution to entropy is negligible). For the martensite to austenite transformation (reverse transformation) in metamagnetic shape memory alloys, the vibrational and the magnetic terms are positive and negative, respectively, the vibrational one being necessarily higher [44]. The opposite occurs in the forward MT being the vibration entropy change negative and positive the magnetic contribution. The MT occurs around the Debye temperature (θ_D = 310 ± 25 K for austenite in a similar but ordered alloy [45]), and then a nearly constant vibrational contribution in a first approximation can be assume in both forward and reverse MT. Since ΔS_{mag} is directly related to ΔM, a higher magnetic contribution opposing the vibrational one should be expected in the forward (lower temperature) than in the reverse transformation (closer to the Curie temperature), which would lead to a lower net $\Delta S^{O}{}_{f}$ in the forward MT than in the reverse $\Delta S^{O}{}_{r}$. So, the difference (in absolute values) between forward and reverse entropies should be always negative. The inset in Figure 1 right shows the evolution of this difference (normalized to the forward entropy change to avoid volume transforming effects) and only for short milling times the difference is negative. In fact the difference turns into positive for milling times longer than 15 min. The large contribution of the magnetic entropy and its temperature dependence makes this system not easy to analyze. In particular, to analyze the effect of defects on the entropy change the previous contributions should be well stablished. ΔS typically increases because of the microstructural recovery brought by annealing, as in fact occurs in ternary Ni-Mn-Sn alloys [46]). This indicates that defects tend to reduce the entropy change linked to the MT. Nevertheless, the influence of defects (point defects, dislocations, and anti-phase boundaries) in the MT thermodynamics should be different in the direct and reverse transformation. The interphase creation and its movement during the MT produces frictional contributions to the entropy change (ΔS_F) both during forward and reverse transformation. In both cases, the entropy change term is negative (exothermic process). So, taking into account this contribution, the entropy change can be written as: ΔS_f $(-) \approx \Delta S^{O}{}_{f}$ $(-) + \Delta S_{F,f}$ $(-)$ for the direct and ΔS_r $(+) \approx \Delta S^{O}{}_{r}$ $(+) + \Delta S_{F,r}$ $(-)$ for the reverse. As a consequence, the absolute value ΔS_f increases and ΔS_r decreases. So, long milling times makes the difference ($\Delta S_f - \Delta S_r$) positive in absolute values (see inset in Figure 1 right). Therefore, defects influence on the entropy change through the frictional terms and then modifies the magnetocaloric capacities and all the other properties linked to the thermodynamics of the MT.

The crystallographic analysis performed after different milling times allow to determine the phase distribution at high (350 K) and low (245 K) temperatures (see Figure 2). The bulk alloy at high temperature shows an austenitic structure (not shown here) but after hand crushing, the induced stresses prompt a small partial transformation to martensite. The subsequent ball milling process induces the presence of much higher martensite fractions at high temperatures (see Figure 2, left). This effect is well known and is usually linked to stress-induced martensitic transformations [47–49]. Besides, an increasing fraction of amorphous phase seems to be present as deformation increase. On heating, the transformation to austenite occurs but the required temperature depends on the stabilization degree. In the present case, the diffraction spectra were measured at 350 K after heating the sample to 420 K indicating that some martensite is still stabilized at 420 K. On the other side, at low temperatures, the presence of higher martensite fractions shows that the MT is taking place. The evolution of the austenite reflections (see reflections at $2\theta = 7.4°$ or $21.2°$ i.e.,) shows that some austenite is retained at low temperatures because of deformation. Nevertheless, the higher is the deformation (longer milling times) the lower is the quantity of out-of-equilibrium austenite at low temperatures (the fraction of austenite at 250 K change from 15% in the As-crushed and 30′milled samples sample to 8% in the sample milled during 60′). Moreover, there exists an increasing amorphous contribution (broad peaks superimposed to the main reflections) with the milling times.

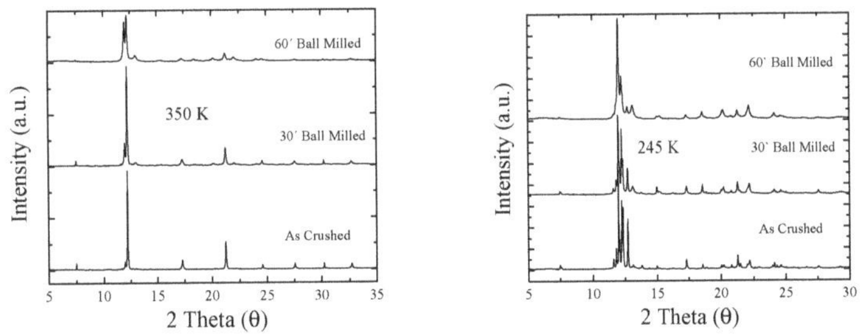

Figure 2. Powder X-ray diffraction patterns at 350 K (**left**) and 245 K (**right**) for the as-crushed and 30´and 60´milled samples. The measurements were performed after cooling from 420 K.

The gradual widening of the reflection peaks with the milling time evidences the expected decrease of the grain size and increase of internal strains. Figure 3 shows the evolution of both parameters estimated from Lebail fit of the 420 K diffractograms obtained for milling times up to 60´. For longer times, the MT practically disappears. So, deformation induces internal stresses that promotes the martensite formation. The percentage of sample undergoing the MT highly decreases with the milling time since the induced martensite does not transform on cooling. This agrees with the reduction of the enthalpy change at MT with the milling time (see Figure 1 right).

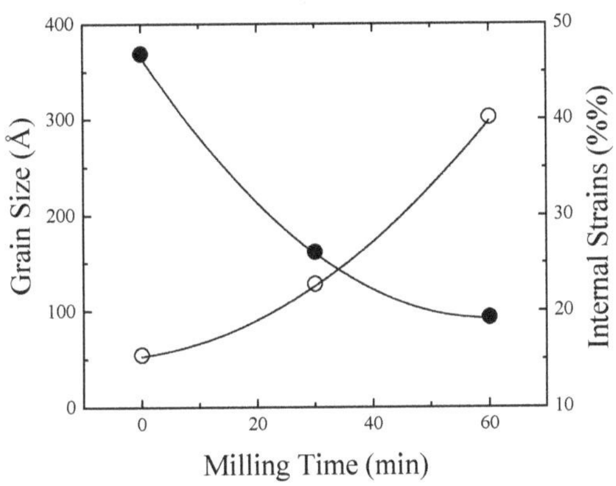

Figure 3. Evolution of the diffracting size (black holes) and internal strains (white holes) measured at 420 K as a function of the milling time.

Magnetic measurements have been carried out on the as-crushed and milled samples to compare the microstructure with the magnetic behavior. Figure 4 shows the temperature dependence of the magnetization at high (60 kOe) and low (100 Oe) magnetic fields. The ferromagnetic austenite transforms to a superparamagnetic-like martensite, showing the characteristic thermal hysteresis. Obviously, the induction of superparamagnetic-like martensite during milling reduces the high field magnetization above the MT. For the longer milling time, the MT is hardly detectable. So, the jump in high-field magnetization at the MT decreases with the milling time, as expected from the gradual increase of the stress-induced martensite. On the other side, the magnetization jump at low magnetic fields linked to the Curie temperature occurs at around 400 K except in the sample milled more than 60 min. In all cases, the MT temperature shows always approximately the same shift to lower temperatures upon the application of high magnetic fields. Besides, both the Curie temperature and the MT temperature remain almost constant irrespectively of the milling time, which again would indicate the absence of variations on the long-range atomic order. As shown in Figure 4 (left), there exists a much stronger influence of the mechanically-induced defects on the magnetization in austenite than in martensite. Apart from the magnetic nature of each phase, local atomic order variations at antiphase boundaries (APB) produced between superdislocations as a consequence of the mechanical deformations could be also playing a role [39].

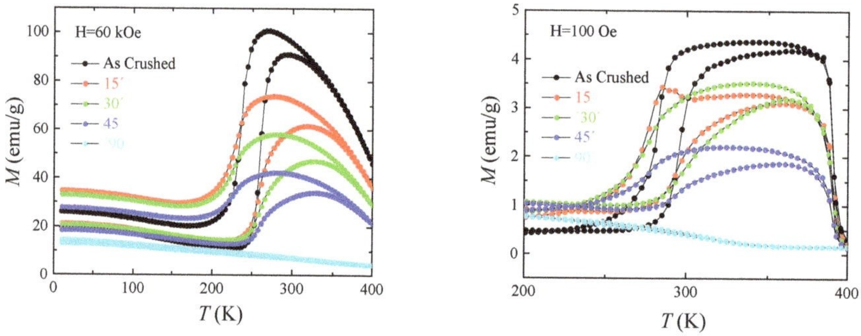

Figure 4. Temperature dependence of the magnetization at high, 60 kOe (**left**) and low, 100 Oe (**right**) magnetic fields.

The magnetic field dependence of the magnetization has been also measured at 10 K to better analyze the effect of milling in martensite. The obtained magnetization curves are shown in Figure 5 (left). For milling times below 30 min, the magnetization curves are similar whereas for longer milling times the magnetization decreases. This result agree with the evolution of the austenite at low temperature; the fraction of ferromagnetic austenite changes from 15% in the as-crushed and 30′milled samples to 8% in the sample milled during 60′. The high magnetization values (20 emu/g) observed in the samples milled during short milling-times is a consequence of the high retained fraction of austenite. Besides, the presence of increasing amounts of amorphous phase with apparently very low magnetization and the reduction in the content of austenite for longer milling times reduces the magnetization of the alloy.

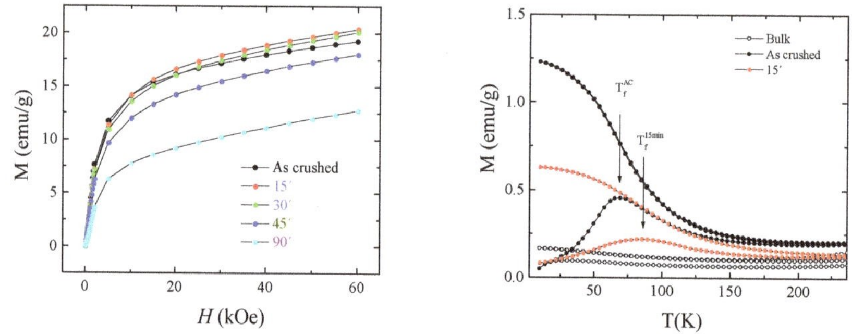

Figure 5. Left: Magnetization *M*(*H*) curves at 10 K after a Zero Field Cooled procedure (ZFC). **Right**: ZFC-Field Cooled (FC) measurement under low magnetic field (100 Oe).

Figure 5 right shows the magnetization versus temperature at 100 Oe without deformation (bulk) and with deformation (as-crushed and 15′ ball milling). The thermomagnetic curves were measured by two processes; cooled to 10 K in zero field and then measured on heating (ZFC curve), and measurement on subsequent cooling (FC curve). The difference between both curves at low temperatures in both deformed samples shows the thermomagnetic history effect typical of the spin-glass behaviour [50]. The maximum in the curves corresponds to the freezing temperature, T_f [50]. The bulk sample does not show the presence of such a maximum. Deformation makes the magnetization (the magnetic susceptibility) increase but the milled sample shows lower susceptibility than the as-crushed sample.

The low magnetic moment of the martensite (see Figure 4 left) must be associated to the coexistence of ferromagnetic and antiferromagnetic interactions. Different mechanisms, i.e., reentrant spin glasses, superparamagnetism, superspin glasses or antiferromagnetism have been proposed to explain their superparamagnetic-like behavior [50–52]. In particular, the presence of local changes in the interatomic distance has been shown to promote changes in the magnetic interaction. This occurs in the Antiphase Boundaries (APB) region between superdislocations in ordered alloys [51]. So, a deformation dependence of the low field magnetization could be expected. APB´s can be considered as regions of enhanced ferromagnetic interaction that increase the overall susceptibility of the phase. The presence of smaller grains and higher strains in the milled sample (see Figure 3) can reduce the overall susceptibility with respect to the as-crushed alloy. Nevertheless, this point requires a more exhaustive analysis. On the other side, the freezing temperature increases slightly with increasing strain (see Figure 5). The shift of the freezing temperature to higher temperature as a consequence of the amount of defects has been observed in other materials [53]. Before deformation, the paramagnetic state does not show any spin-glass like behavior. Indeed, the deformed samples have two magnetic states above T_f, ferromagnetic clusters locate along the APB and paramagnetic-like state surrounding the APB. These paramagnetic and ferromagnetic states change to the spin-glass at T_f in the deformed samples. The results indicate that the freezing temperature T_f depends on the deformation degree, increasing with the quantity of dislocations [51].

Although other approaches have been proposed in the literature based on the temperature derivative of the Helmoltz-free energy [54,55], the MCE, defined as the field induced entropy change ΔS_{iso} in isothermal conditions, has been calculated according to classical thermodynamics through the temperature dependence of magnetization (curves not shown here but similar to those shown in Figure 4 but at different applied magnetic fields). The measured ΔS_{iso} for a powder milled during 15´ as a function of temperature and magnetic field is shown in Figure 6.

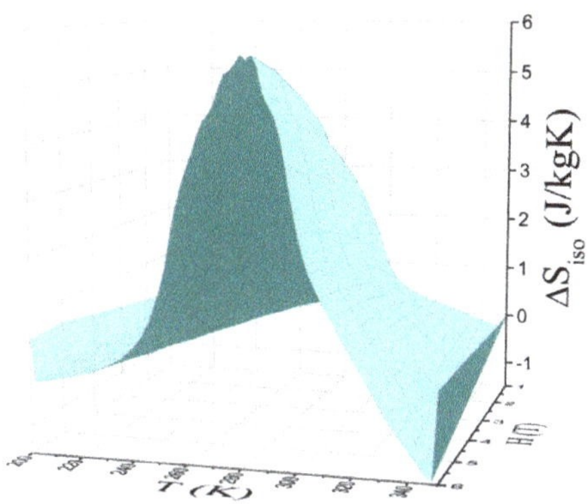

Figure 6. Temperature and field dependence of the magnetic field induced entropy change (ΔS) in a 15´ ball milled sample.

A positive peak (inverse MCE) is observed linked to the magnetostructural transformation at the MT. The MCE values increase with the increasing magnetic field, reaching a maximum value of around $\Delta S_{iso} \approx 5$ J/kgK at 60 kOe. The ΔS_{iso} comes from the magnetic induction of the MT which is limited by the maximum entropy change found in Figure 2 for the sample milled during 15´ ($\Delta S \approx 5$ J/kgK). The results seem to indicate that a field of 60 kOe is high enough to induce the full available MT. Although the measured value is small comparing with the bulk entropy change, a subsequent high temperature thermal treatment should remove the defects and improve their MCE properties. Further work is in progress to improve their MCE capabilities.

From the application point of view, the production of MSMA μ-particles opens the possibility of producing MSMA-particle containing polymers (composites) by different techniques and in particular by Additive Manufacturing (AM) technology (3D printing). The new AM technology opens a huge field, for example, in the design of heat exchangers for magnetic refrigeration grounded on the inverse magnetocaloric effect or in the design of tunable (using for example external magnetic fields) damping devices. To check the potential capacities in both fields, preliminary studies have been performed on the particles, concerning the magnetic field induced entropy change and the damping properties.

For this last application, the hand crushed powders were embedded in a photosensitive polymer. Figure 7 shows the temperature dependence of the damping ($\tan(\phi)$) and dynamic shear modulus (G') of the composite. The damping spectra show the peak related to the glass transition of the polymer, Tg, at around 340 K and its related fall in the dynamic modulus at around 300 K, overlapped to the MT. Nevertheless, composites show a well-defined damping peak at around 285 K (labelled as MT), overlapped on the low temperature side of the glass transition peak. There exists a clear change in the shear modulus linked to the damping peak. Both the peak and the modulus defect show a temperature hysteresis related to the MT. So, the damping of the polymer reflects the effect of the MT and consequently the application of external magnetic fields could change for example the transformation temperature and consequently the temperature range where the damping is maximum. Obviously, much more efforts are necessary to understand and design tunable dampers using this technology but these previous results seem to be promising.

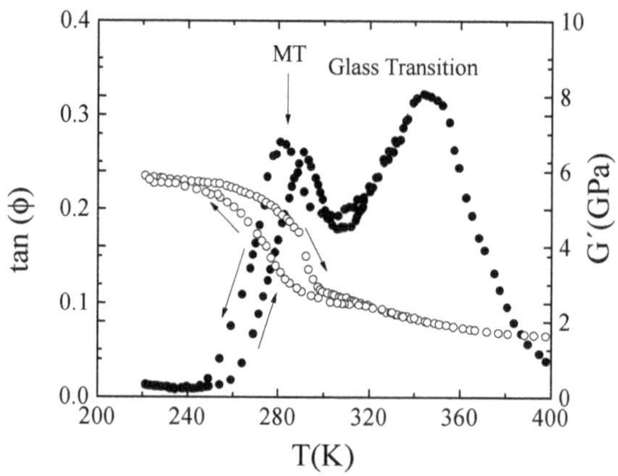

Figure 7. Damping ($\tan(\phi)$, black holes) and shear modulus (G´, white holes) as function of temperature measured in a Metamagnetic Shape Memory Alloy-Polymer composite.

4. Conclusions

A crushing and subsequent ball milling process has been tested to produce MSMA μ-particles. The particles in the deformed state were introduced in a photosensitive polymer to check their damping capacity. The high energy involved in the milling process produces large internal stresses and high density of defects with a strong influence on the MT characteristics. Deformations induce internal stresses that promotes the MT; the interphase creation and its movement during the MT produces frictional contributions to the entropy change (exothermic process) both during forward and reverse transformation modifying the absolute values of the total entropy. The frictional contribution increases with the milling time as a consequence of the interaction between defects and interphases. The influence of the frictional terms on the MCE has been evidenced. Besides, the presence of antiphase boundaries linked to superdislocations helps to understand the spin-glass behavior at low temperatures in martensite. Finally, the damping of the produced composites reflects the MT, so the elaborated composites appear to be promising candidates to the development of mechanical dampers tuned by external magnetic fields.

Author Contributions: Conceptualization, J.I.P.-L., V.S.-A. and V.R.; methodology, J.I.P.-L., V.S.-A. and V.R.; formal analysis J.I.P.-L., O.A.L., J.A.G. and F.P.; investigation, I.U., J.L.-G., M.J.R., J.A.R.-V. and E.C.; writing—original draft preparation, J.I.P.-L.; writing—review and editing, J.I.P.-L., V.S.-A., V.R., O.A.L., F.G.B., D.L.R.K., I.U., J.A.G., F.P., J.L.-G., M.J.R., J.A.R.-V. and E.C.; supervision, J.I.P.-L.; All authors have read and agreed to the published version of the manuscript.

Funding: This research was funded by Projects RTI2018-094683-B-C5 (4,5) (MCIU/AEI/FEDER,UE); ASACTEI Pcia.Santa Fe IO-2017-00138, PID-UNR ING 575 and ING 612 (2018–2021).

Acknowledgments: D.L.R.K. wants to acknowledge UPNA and InaMat^2 for the Ph.D. Grant (Ayudas para contratos pre-doctorales adscritas a grupos e instittutos de investigación UPNA) We acknowledge ALBA Synchrotron for the beam time allocations: BL04_MPSD beamline with the collaboration of ALBA staff (Proposal ref. 2018022796). Authors also want to thank the Cooperation Agreement between the Universidad Pública de Navarra and the Universidad Nacional de Rosario, Res. CS. 3247/2015. 2015-2020 and the Cooperation Agreement between the Universidad del País Vasco and the Universidad Nacional de Rosario, Res. C. S. 3243 del 30/12/2015. 2015–2020.

Conflicts of Interest: The authors declare no conflict of interest.

References

1. Kainuma, R.; Imano, Y.; Ito, W.; Sutou, Y.; Morito, H.; Okamoto, S.; Kitakami, O.; Oikawa, K.; Fujita, A.; Kanomata, T.; et al. Magnetic-field-induced shape recovery by reverse phase transformation. *Nature* **2006**, *439*, 957–960. [CrossRef] [PubMed]
2. Koyama, K.; Watanabe, K.; Kanomata, T.; Kainuma, R.; Oikawa, K.; Ishida, K. Observation of field-induced reverse transformation in ferromagnetic shape memory alloy Ni50Mn36Sn14. *Appl. Phys. Lett.* **2006**, *88*, 132505. [CrossRef]
3. Oikawa, K.; Ito, W.; Imano, Y.; Sutou, Y.; Kainuma, R.; Ishida, K.; Okamoto, S.; Kitakami, O.; Kanomata, T. Effect of magnetic field on martensitic transition of Ni46Mn41In13 Heusler alloy. *Appl. Phys. Lett.* **2006**, *88*, 122507. [CrossRef]
4. Acet, M.; Mañosa, L.; Planes, A. *Chapter Four—Magnetic-Field-Induced Effects in Martensitic Heusler-Based Magnetic Shape Memory Alloys*; Buschow, K.H.J., Ed.; Elsevier: Amsterdam, The Netherlands, 2011; Volume 19.
5. Şaşıoğlu, E.; Sandratskii, L.M.; Bruno, P. First-principles calculation of the intersublattice exchange interactions and Curie temperatures of the full Heusler alloys Ni2MnX (X=Ga, In, Sn). *Phys. Rev. B* **2004**, *70*, 24427. [CrossRef]
6. Şaşıoğlu, E.; Sandratskii, L.M.; Bruno, P. Role of conduction electrons in mediating exchange interactions in Mn-based Heusler alloys. *Phys. Rev. B* **2008**, *77*, 64417. [CrossRef]
7. Buchelnikov, V.D.; Entel, P.; Taskaev, S.V.; Sokolovskiy, V.V.; Hucht, A.; Ogura, M.; Akai, H.; Gruner, M.E.; Nayak, S.K. Monte Carlo study of the influence of antiferromagnetic exchange interactions on the phase transitions of ferromagnetic Ni-Mn-X alloys (X=In, Sn, Sb). *Phys. Rev. B* **2008**, *78*, 184427. [CrossRef]
8. Kainuma, R.; Ito, W.; Umetsu, R.Y.; Khovaylo, V.V.; Kanomata, T. Metamagnetic shape memory effect and magnetic properties of Ni-Mn based Heusler alloys. *Mater. Sci. Forum* **2011**, *684*, 139–150. [CrossRef]
9. Yu, S.Y.; Liu, Z.H.; Liu, G.D.; Chen, J.L.; Cao, Z.X.; Wu, G.H.; Zhang, B.; Zhang, X.X. Large magnetoresistance in single-crystalline Ni50Mn50−xInx alloys (x=14–16) upon martensitic transformation. *Appl. Phys. Lett.* **2006**, *89*, 162503. [CrossRef]
10. Sharma, V.K.; Chattopadhyay, M.K.; Shaeb, K.H.B.; Chouhan, A.; Roy, S.B. Large magnetoresistance in Ni50Mn34In16 alloy. *Appl. Phys. Lett.* **2006**, *89*, 222509. [CrossRef]
11. Krenke, T.; Duman, E.; Acet, M.; Wassermann, E.F.; Moya, X.; Mañosa, L.; Planes, A. Inverse magnetocaloric effect in ferromagnetic Ni-Mn-Sn alloys. *Nat. Mater.* **2005**, *4*, 450–454. [CrossRef]
12. Krenke, T.; Duman, E.; Acet, M.; Wassermann, E.F.; Moya, X.; Mañosa, L.; Planes, A.; Suard, E.; Ouladdiaf, B. Magnetic superelasticity and inverse magnetocaloric effect in Ni-Mn-In. *Phys. Rev. B* **2007**, *75*, 104414. [CrossRef]
13. Planes, A.; Mañosa, L.; Acet, M. Magnetocaloric effect and its relation to shape-memory properties in ferromagnetic Heusler alloys. *J. Phys. Condens. Matter* **2009**, *21*, 233201. [CrossRef] [PubMed]
14. Liu, J.; Gottschall, T.; Skokov, K.P.; Moore, J.D.; Gutfleisch, O. Giant magnetocaloric effect driven by structural transitions. *Nat. Mater.* **2012**, *11*, 620–626. [CrossRef] [PubMed]

15. Pérez-Landazábal, J.I.; Lambri, O.A.; Bonifacich, F.G.; Sánchez-Alarcos, V.; Recarte, V.; Tarditti, F. Influence of defects on the irreversible phase transition in Fe–Pd ferromagnetic shape memory alloys. *Acta Mater.* **2015**, *86*, 110–117. [CrossRef]
16. Bonifacich, F.G.; Lambri, O.A.; Pérez-Landazábal, J.I.; Gargicevich, D.; Recarte, V.; Sánchez-Alarcos, V. Mobility of Twin Boundaries in Fe-Pd-Based Ferromagnetic Shape Memory Alloys. *Mater. Trans.* **2016**, *57*, 1837–1844. [CrossRef]
17. Sutou, Y.; Imano, Y.; Koeda, N.; Omori, T.; Kainuma, R.; Ishida, K.; Oikawa, K. Magnetic and martensitic transformations of NiMnX(X=In,Sn,Sb) ferromagnetic shape memory alloys. *Appl. Phys. Lett.* **2004**, *85*, 4358–4360. [CrossRef]
18. Krenke, T.; Acet, M.; Wassermann, E.F.; Moya, X.; Mañosa, L.; Planes, A. Martensitic transitions and the nature of ferromagnetism in the austenitic and martensitic states of Ni-Mn-Sn alloys. *Phys. Rev. B* **2005**, *72*, 14412. [CrossRef]
19. Krenke, T.; Acet, M.; Wassermann, E.F.; Moya, X.; Mañosa, L.; Planes, A. Ferromagnetism in the austenitic and martensitic states of Ni-Mn-In alloys. *Phys. Rev. B* **2006**, *73*, 174413. [CrossRef]
20. Khan, M.; Dubenko, I.; Stadler, S.; Ali, N. Magnetostructural phase transitions in Ni50Mn25+xSb25−xHeusler alloys. *J. Phys. Condens. Matter* **2008**, *20*, 235204. [CrossRef]
21. Recarte, V.; Pérez-Landazábal, J.I.; Sánchez-Alarcos, V.; Rodríguez-Velamazán, J.A. Dependence of the martensitic transformation and magnetic transition on the atomic order in Ni–Mn–In metamagnetic shape memory alloys. *Acta Mater.* **2012**, *60*, 1937–1945. [CrossRef]
22. Recarte, V.; Pérez-Landazábal, J.I.; Sánchez-Alarcos, V. Dependence of the relative stability between austenite and martensite phases on the atomic order in a Ni–Mn–In Metamagnetic Shape Memory Alloy. *J. Alloys Compd.* **2012**, *536*, S308–S311. [CrossRef]
23. Sánchez-Alarcos, V.; Recarte, V.; Pérez-Landazábal, J.; Cesari, E.; Rodríguez-Velamazán, J. Long-Range Atomic Order and Entropy Change at the Martensitic Transformation in a Ni-Mn-In-Co Metamagnetic Shape Memory Alloy. *Entropy* **2014**, *16*, 2756–2767. [CrossRef]
24. Sánchez-Alarcos, V.; Pérez-Landazábal, J.I.; Recarte, V.; Lucia, I.; Vélez, J.; Rodríguez-Velamazán, J.A. Effect of high-temperature quenching on the magnetostructural transformations and the long-range atomic order of Ni–Mn–Sn and Ni–Mn–Sb metamagnetic shape memory alloys. *Acta Mater.* **2013**, *61*, 4676–4682. [CrossRef]
25. Kainuma, R.; Imano, Y.; Ito, W.; Morito, H.; Sutou, Y.; Oikawa, K.; Fujita, A.; Ishida, K.; Okamoto, S.; Kitakami, O.; et al. Metamagnetic shape memory effect in a Heusler-type Ni43Co7Mn39Sn11 polycrystalline alloy. *Appl. Phys. Lett.* **2006**, *88*, 192513. [CrossRef]
26. Nayak, A.K.; Suresh, K.G.; Nigam, A.K. Giant inverse magnetocaloric effect near room temperature in Co substituted NiMnSb Heusler alloys. *J. Phys. D. Appl. Phys.* **2009**, *42*, 35009. [CrossRef]
27. Huang, L.; Cong, D.; Suo, H.; Wang, Y. Giant magnetic refrigeration capacity near room temperature in Ni40Co10Mn40Sn10 multifunctional alloy. *Appl. Phys. Lett.* **2014**, *104*, 132407. [CrossRef]
28. Umetsu, R.Y.; Sheikh, A.; Ito, W.; Ouladdiaf, B.; Ziebeck, K.R.A.; Kanomata, T.; Kainuma, R. The effect of Co substitution on the magnetic properties of the Heusler alloy Ni50Mn33Sn17. *Appl. Phys. Lett.* **2011**, *98*, 42507. [CrossRef]
29. Lahelin, M.; Aaltio, I.; Heczko, O.; Söderberg, O.; Ge, Y.; Löfgren, B.; Hannula, S.-P.; Seppälä, J. DMA testing of Ni–Mn–Ga/polymer composites. *Compos. Part A Appl. Sci. Manuf.* **2009**, *40*, 125–129. [CrossRef]
30. Scheerbaum, N.; Hinz, D.; Gutfleisch, O.; Müller, K.-H.; Schultz, L. Textured polymer bonded composites with Ni–Mn–Ga magnetic shape memory particles. *Acta Mater.* **2007**, *55*, 2707–2713. [CrossRef]
31. Feuchtwanger, J.; Richard, M.L.; Tang, Y.J.; Berkowitz, A.E.; O'Handley, R.C.; Allen, S.M. Large energy absorption in Ni–Mn–Ga/polymer composites. *J. Appl. Phys.* **2005**, *97*, 10M319. [CrossRef]
32. Liu, J.; Scheerbaum, N.; Kauffmann-Weiss, S.; Gutfleisch, O. NiMn-Based Alloys and Composites for Magnetically Controlled Dampers and Actuators. *Adv. Eng. Mater.* **2012**, *14*, 653–667. [CrossRef]
33. Tian, B.; Chen, F.; Tong, Y.; Li, L.; Zheng, Y. Magnetic field induced strain and damping behavior of Ni–Mn–Ga particles/epoxy resin composite. *J. Alloys Compd.* **2014**, *604*, 137–141. [CrossRef]
34. Alves, A.L.; Passamani, E.C.; Nascimento, V.P.; Takeuchi, A.Y.; Larica, C. Influence of grain refinement and induced crystal defects on the magnetic properties of Ni50Mn36Sn14Heusler alloy. *J. Phys. D Appl. Phys.* **2010**, *43*, 345001. [CrossRef]

35. Passamani, E.; Nascimento, V.; Larica, C.; Takeuchi, A.; Alves, A.; Capua Proveti, J.; Pereira, M.; Fabris, J. The influence of chemical disorder enhancement on the martensitic transformation of the Ni 50Mn 36Sn 14 Heusler-type alloy. *J. Alloy. Compd.* **2011**, *509*, 7826–7832. [CrossRef]
36. Ghotbi Varzaneh, A.; Kameli, P.; Zahedi, V.R.; Karimzadeh, F.; Salamati, H. Effect of heat treatment on martensitic transformation of Ni47Mn40Sn13 ferromagnetic shape memory alloy prepared by mechanical alloying. *Met. Mater. Int.* **2015**, *21*, 758–764. [CrossRef]
37. Czaja, P.; Przewoźnik, J.; Fitta, M.; Bałanda, M.; Chrobak, A.; Kania, B.; Zackiewicz, P.; Wójcik, A.; Szlezynger, M.; Maziarz, W. Effect of ball milling and thermal treatment on exchange bias and magnetocaloric properties of Ni48Mn39.5Sn10.5Al2 ribbons. *J. Magn. Magn. Mater.* **2016**, *401*, 223–230. [CrossRef]
38. Wang, X.; Sun, F.; Wang, J.; Yu, Q.; Wu, Y.; Hua, H.; Jiang, C. Influence of annealing temperatures on the magnetostructural transition and magnetocaloric effect of Ni40Co10Mn40Sn10 powders. *J. Alloys Compd.* **2017**, *691*, 215–219. [CrossRef]
39. Unzueta, I.; López-García, J.; Sánchez-Alarcos, V.; Recarte, V.; Pérez-Landazábal, J.I.; Rodríguez-Velamazán, J.A.; Garitaonandia, J.S.; García, J.A.; Plazaola, F. 119Sn Mössbauer spectroscopy for assessing the local stress and defect state towards the tuning of Ni-Mn-Sn alloys. *Appl. Phys. Lett.* **2017**, *110*, 181908. [CrossRef]
40. Murakami, Y.; Yanagisawa, K.; Niitsu, K.; Park, H.S.; Matsuda, T.; Kainuma, R.; Shindo, D.; Tonomura, A. Determination of magnetic flux density at the nanometer-scale antiphase boundary in Heusler alloy Ni50Mn25Al12.5Ga12.5. *Acta Mater.* **2013**, *61*, 2095–2101. [CrossRef]
41. Fauth, F.; Peral, I.; Popescu, C.; Knapp, M. The new Material Science Powder Diffraction beamline at ALBA Synchrotron. *Powder Diffr.* **2013**, *28*, S360–S370. [CrossRef]
42. Peral, I.; McKinlay, J.; Knapp, M.; Ferrer, S. Design and construction of multicrystal analyser detectors using Rowland circles: Application to MAD26 at ALBA. *J. Synchrotron Radiat.* **2011**, *18*, 842–850. [CrossRef] [PubMed]
43. Rodríguez-Carvajal, J. Recent advances in magnetic structure determination by neutron powder diffraction. *Phys. B Condens. Matter* **1993**, *192*, 55–69. [CrossRef]
44. Recarte, V.; Pérez-Landazábal, J.I.; Sánchez-Alarcos, V.; Zablotskii, V.; Cesari, E.; Kustov, S. Entropy change linked to the martensitic transformation in metamagnetic shape memory alloys. *Acta Mater.* **2012**, *60*, 3168–3175. [CrossRef]
45. Pérez-Landazábal, J.I.; Recarte, V.; Sánchez-Alarcos, V.; Ruiz, M.J.; Cesari, E. Outstanding role of the magnetic entropy in arrested austenite in an ordered Ni45Mn36.7In13.3Co5 metamagnetic shape memory alloy. *Scr. Mater.* **2019**, *168*, 91–95. [CrossRef]
46. López-García, J.; Unzueta, I.; Sánchez-Alarcos, V.; Recarte, V.; Pérez-Landazábal, J.I.; Rodríguez-Velamazán, J.A.; García, J.A.; Plazaola, F. Correlation between defects and magneto-structural properties in Ni-Mn-Sn metamagnetic shape memory alloys. *Intermetallics* **2018**, *94*, 133–137. [CrossRef]
47. Li, G.T.; Liu, Z.H.; Ma, X.Q.; Yu, S.Y.; Liu, Y. Grinding-induced martensite stabilization in Mn50Ni33.5Sn8Co8.5 alloy. *Mater. Lett.* **2013**, *107*, 239–242. [CrossRef]
48. Singh, S.; Kushwaha, P.; Scheibel, F.; Liermann, H.-P.; Barman, S.R.; Acet, M.; Felser, C.; Pandey, D. Residual stress induced stabilization of martensite phase and its effect on the magnetostructural transition in Mn-rich Ni-Mn-In/Ga magnetic shape-memory alloys. *Phys. Rev. B* **2015**, *92*, 20105. [CrossRef]
49. Chulist, R.; Czaja, P.; Tokarski, T.; Faryna, M. Martensite stabilisation in single crystalline Ni-Mn-Ga and Ni-Mn-Sn magnetic shape memory alloys. *Mater. Lett.* **2018**, *230*, 266–269. [CrossRef]
50. Umetsu, R.Y.; Fujita, A.; Ito, W.; Kanomata, T.; Kainuma, R. Determination of the magnetic ground state in the martensite phase of Ni–Mn–Z (Z= In, Sn and Sb) off-stoichiometric Heusler alloys by nonlinear AC susceptibility. *J. Phys. Condens. Matter* **2011**, *23*, 326001. [CrossRef]
51. Larumbe, S.; Unzueta, I.; Sánchez-Alarcos, V.; Pérez-Landazábal, J.I.; Recarte, V.; García, J.A.; Plazaola, F. Low temperature magnetic properties of a Ni50Mn34In16 ball-milled metamagnetic shape memory alloy. *J. Non. Cryst. Solids* **2016**, *447*, 16–20. [CrossRef]
52. Sánchez-Alarcos, V.; Recarte, V.; Pérez-Landazábal, J.I.; Larumbe, S.; Caballero-Flores, R.; Unzueta, I.; García, J.A.; Plazaola, F.; Rodríguez-Velamazán, J.A. Mechanically induced disorder and crystallization process in Ni-Mn-In ball-milled alloys. *J. Alloys Compd.* **2016**, *689*, 983–991. [CrossRef]
53. Takahashi, S.; Chiba, A.; Takahashi, E. The influence of plastic deformation on the spin-glass in Fe-37.0 at.% Al compound. *Phys. Lett. A* **1995**, *197*, 350–352. [CrossRef]

54. L'vov, V.A.; Kosogor, A.; Barandiaran, J.M.; Chernenko, V.A. Theoretical description of magnetocaloric effect in the shape memory alloy exhibiting metamagnetic behavior. *J. Appl. Phys.* **2016**, *119*, 13902. [CrossRef]
55. Kosogor, A.; L'vov, V.; Lázpita, P.; Seguí, C.; Cesari, E. Magnetocaloric Effect Caused by Paramagnetic Austenite–Ferromagnetic Martensite Phase Transformation. *Metals* **2018**, *9*, 11. [CrossRef]

metals

MDPI

Article

Laser Powder Bed Fusion Processing of Fe-Mn-Al-Ni Shape Memory Alloy—On the Effect of Elevated Platform Temperatures

Felix Clemens Ewald [1], Florian Brenne [1], Tobias Gustmann [2], Malte Vollmer [1], Philipp Krooß [1,*] and Thomas Niendorf [1]

[1] Institute of Materials Engineering—Metallic Alloys, University of Kassel, 34125 Kassel, Germany; Ewald@uni-kassel.de (F.C.E.); florian.brenne@gmail.com (F.B.); vollmer@uni-kassel.de (M.V.); niendorf@uni-kassel.de (T.N.)

[2] Leibniz IFW Dresden, Institute for Complex Materials, 01069 Dresden, Germany; t.gustmann@ifw-dresden.de

* Correspondence: krooss@uni-kassel.de

Abstract: In order to overcome constraints related to crack formation during additive processing (laser powder bed fusion, L-BPF) of Fe-Mn-Al-Ni, the potential of high-temperature L-PBF processing was investigated in the present study. The effect of the process parameters on crack formation, grain structure, and phase distribution in the as-built condition, as well as in the course of cyclic heat treatment was examined by microstructural analysis. Optimized processing parameters were applied to fabricate cylindrical samples featuring a crack-free and columnar grained microstructure. In the course of cyclic heat treatment, abnormal grain growth (AGG) sets in, eventually promoting the evolution of a bamboo like microstructure. Testing under tensile load revealed a well-defined stress plateau and reversible strains of up to 4%.

Keywords: additive manufacturing; laser powder bed fusion; shape memory alloy; Fe-Mn-Al-Ni; cyclic heat treatment

Citation: Ewald, F.C.; Brenne, F.; Gustmann, T.; Vollmer, M.; Krooß, P.; Niendorf, T. Laser Powder Bed Fusion Processing of Fe-Mn-Al-Ni Shape Memory Alloy—On the Effect of Elevated Platform Temperatures. *Metals* **2021**, *11*, 185. https://doi.org/10.3390/met11020185

Academic Editor: Gabriel A. Lopez

Received: 27 December 2020
Accepted: 15 January 2021
Published: 20 January 2021

1. Introduction

During the past decades, the development of new shape memory alloys (SMAs) gained a lot of interest in academia and the industry, in order to meet the demand for light and extremely compact actuators and damping devices. It is well-known that binary Ni-Ti SMAs suffer from poor cold workability, relatively high costs, as well as a narrow temperature window for practical application, limiting a more widespread industrial usage [1]. Thus, other SMA systems came into focus of academic and industrial research to extend the application range, e.g., to higher transformation temperatures, as in the case of Ni-Ti-Hf alloys [2], or to higher application frequencies, as in the case of magnetic SMAs [3,4]. For providing more economic, low-cost SMAs, substantial work was conducted on iron-based SMAs, such as Fe-Mn-Si [5], Fe-Ni-Co-Al-Ta [6], and Fe-Mn-Al-Ni [7]. Based on these alloys it is possible to realize novel applications, e.g., in the construction sector [8].

However, good functional properties strongly require adequate microstructures in Fe-based SMAs. Thus, the microstructure should be tailored with respect to the crystallographic orientation, grain size, grain morphology, and phase fractions, in order to enable high transformation strains, good reversibility, as well as cyclic stability [5,9–17]. Thus, complex thermo-mechanical treatments like combined hot-cold rolling, wire drawing, and cyclic heat treatments were investigated in several studies [5,9–17]. However, even if there is reasonable progress, efforts spent on thermomechanical processing are tremendous, and up to now microstructure design in Fe-based SMAs still remains challenging.

With the emergence of the additive manufacturing (AM) technologies, such as laser powder bed fusion (L-PBF) or electron beam powder bed fusion (E-PBF), which enable the production of components, layer by layer, on the basis of a computer-aided design (CAD) file, new possibilities are opened up. Based on this principle, a high geometric complexity is offered without the need of tooling operations [18]. Furthermore, via L-PBF and E-PBF the microstructure with respect to grain size, grain morphology, and texture can be tailored by applying different exposure strategies for melting the initial powdery starting material, e.g., as shown for Ni-base alloys [19], aluminum [20], and stainless steel 316 L [21,22]. In consequence, processing of SMAs by AM techniques came into focus of research [23–26]. Numerous studies focused on AM of Ni-Ti revealed that the processing window of Ni-Ti seems to be rather small [27]. Recently, some of the current authors investigated AM of Fe-Mn-Al-Ni by L-PBF [28] as well as E-PBF [29]. It was shown that the alloy is generally suitable for processing by both AM techniques. Adequate functional properties were revealed after additional heat treatments. These post treatments aimed at promoting abnormal grain growth (AGG) [16,30] and precipitate formation [31]. However, up to now, superelastic properties were only shown under compressive loading. Most detrimentally, crack formation occurred especially in material processed by L-PBF. Since it is well known that rapid quenching and cooling, respectively, can lead to crack formation in Fe-Mn-Al-Ni [17], it is likely that cracking occurs due to the high cooling rates and the general thermal history prevailing in L-PBF during processing. To substantiate this assumption, in the present study, Fe-Mn-Al-Ni was processed using a built platform heated up to 500 °C during processing, in order to ensure lower cooling rates during processing and, eventually, promote crack-free conditions, showing superelastic properties in tension.

2. Materials and Methods

Commercially pure Fe-Mn-Al-Ni, produced by vacuum induction melting, with a nominal chemical composition of Fe-34-Mn-14-Al-7.5-Ni (at.%), i.e., Fe 48.1-Mn 36.1. Al 7.3 Ni 8.5 (wt.%), was cast by Stahlzentrum Freiberg e.V. (Freiberg, Germany). Subsequently, the material was gas atomized under argon atmosphere by TLS Technik GmbH (Bitterfeld, Germany). The resulting mean particle size (d_{50}) of the powder used for laser powder bed fusion was 40 μm. Using a SLM250^{HL} machine (SLM Solutions Group AG, Lübeck, Germany), cubes of 5 mm × 5 mm × 7 mm, cylinders of 5 mm in diameter, and 40 mm in height, as well as pillars/rods with diameters ranging from 0.5 to 3 mm (height: 7 mm) were fabricated. From the cylinders, the tensile samples were fabricated by electrode discharge machining. The pillars were conventionally separated from the baseplate and used for the evaluation of the geometric/dimensional impact on microstructure evolution. The build platform (in-house development of Leibniz IFW [32], see Figure 1 for details) was heated to 500 °C before processing. A laser power of 200 W, scan speed of 680 mm s^{-1}, hatch distance of 0.12 mm, and a layer thickness of 30 μm, were employed for sample processing (stripe scanning, −90° scan vector rotation). After the build job ended, the baseplate temperature was further held constant for 15 min followed by slow cooling to 400 °C and switching of the heating system. After the aforementioned cooling phase, the sample setup was quickly removed and placed on an iron-cast block to allow faster cooling to room temperature.

Beside the as-built state, selected samples were heat-treated in evacuated quartz tubes. In a first step, a solution heat treatment at 1200 °C for 12 h was followed by water quenching. Afterwards, defined heat treatment cycles (HTC) were applied. Details on the procedure are depicted in Section 3 for clarity. In the present work, three and five HTC were applied to the selected samples, followed by quenching in 80 °C tempered water.

The chemical composition of the powder and the bulk material (L-PBF samples) was analyzed using ICP-OES (Inductively-Coupled Plasma–Optical Emission Spectroscopy, IRIS Intrepid II XUV from Thermo Fischer Scientific, Waltham, MA, USA). Furthermore, the oxygen (LECO TC-436DR—powder and L-PBF) and hydrogen content (HORIBA EMGA 621 W—powder only) were determined in order to exclude contamination of the powder and the L-PBF processed bulk. A minimum of three measurements were performed to obtain the experimental average values and corresponding deviations.

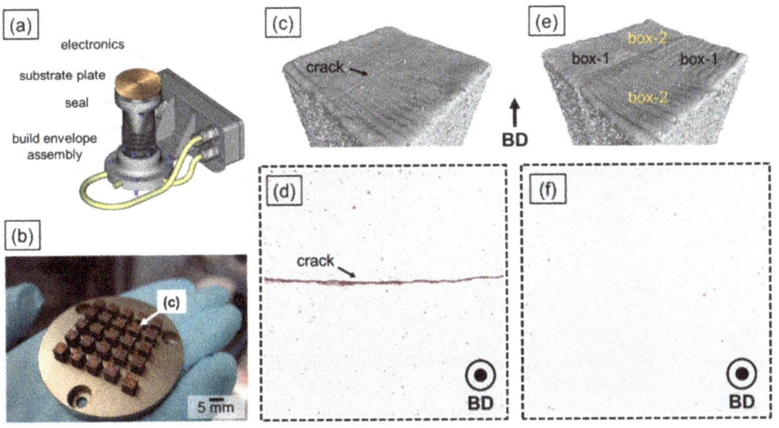

Figure 1. Overview of the machine setup used and selected Fe-Mn-Al-Ni samples fabricated at Leibniz IFW Dresden. An in-house developed build envelope (substrate plate diameter of 67 mm) was used for processing at a fixed temperature of 500 °C (**a**). Process parameter optimization for Fe-Mn-Al-Ni (**b**) considering varying laser power (right to left: 125–225 W) and scanning speed (top to bottom: 610–890 mm/s). Reconstructed CT-images (**c**–**f**) revealing that crack free samples (5 mm × 5 mm × 7 mm) can be manufactured using the optimized parameters laser power = 200 W, scanning speed = 680 mm/s, hatching distance = 0.12 mm, in combination with (**e**,**f**) a chessboard-like scanning pattern (2.5 mm × 2.5 mm × 2 mm box size). Box-1 and box-2, as depicted in (**e**) were scanned perpendicular to each other and were constantly rotated layer-by-layer, by 90°. Numerous pores and a major crack in case of (**c**,**d**) conventional stripe scanning without island separation are shown in red. Please note that the images in (**d**,**f**) show the analyzed, transparent volume of the cubes (top view) shown in (**c**,**e**), respectively.

In order to see if the samples could be processed crack-free and with a low residual porosity, selected samples, fabricated with the optimized process parameters and the high-temperature build envelope, were investigated using a Phoenix Nanotom M (General Electric) computer tomograph (referred to as CT). The resolution of the CT was set to 6 μm, using an acceleration voltage of 135 kV and a current of 100 μA. CT-scans were performed using a beam collimator and copper filter (0.3 mm in thickness). 700 projections were recorded in total for the reconstruction of a single 3D-model. Volume analysis was carried out using VG-Studio max 2.2 (Volume Graphics). Artefacts and imperfections (voids with low sharpness) below the voxel size were digitally filtered out and not further considered.

For scanning electron microscopy (SEM), the samples were mechanically ground, polished down to 5 microns, and finally vibration polished. The SEM used was operated at 20 kV and equipped with an electron backscatter diffraction (EBSD) unit (EDAX Inc, Mahwah, NJ, USA). From EBSD data inverse pole figure (IPF) maps were obtained using standard software tools (EDAX OIM software, version 7, EDAX Inc, Mahwah, NJ, USA). The Vicker's hardness was probed using a load of 4.9 N (HV 0.5). Mechanical testing (controlled loading/unloading cycles) was conducted using flat small dog-bone shaped samples with a gage section of 18 mm, with 1.5mm x 1.5 mm cross section, electro-discharge machined from the as-built cylinders (diameter = 5 mm, height = 40 mm). A servohydraulic testing rig was used at a crosshead displacement speed of 0.01 mm s^{-1}. An extensometer was directly attached to the sample gage section.

3. Results and Discussion

L-BPF processing of the Fe-34Mn-14Al-7.5Ni shape memory alloy at a build platform temperature of 500 °C resulted in a crack-free microstructure in the as-built condition. In Fe-Mn-Al-Ni, several precipitation processes take place as a function of time and temperature [33,34]. From room temperature (RT) up to about 400 °C, small nanometric β-phase precipitates form upon aging, within minutes or hours, depending on the ag ing temperature applied. Niendorf et al. [28] reported that due to the formation of th β-type precipitates and due to the high cooling rates induced by a relatively low buil platform temperature of up to 200 °C, significant crack formation occurs following L-PB processing. It was stated that L-PBF processing at 200 °C leads to a strong increase of th hardness, resulting in an embrittlement of the microstructure. Eventually, the high coolin rates during L-PBF processing and the evolution of internal stress fields led to significar crack formation [28]. In the present study, the platform temperature was fixed at 500 °C Obviously, this had a significant impact on the thermal history and, thus, on the sampl quality and the microstructural evolution. While the thermal gradient during processing i significantly higher in case of a build platform temperature of only 200 °C, a build platforr temperature of 500 °C promotes lower cooling rates, as well as a less pronounced defec formation (cracks, pores, etc.) during processing (Figure 1c–f).

The as-built microstructure was processed and probed in the form of small cylinders o different diameters, in order to evaluate the impact of geometry/dimensions on microstru tural evolution (Figure 2). In the as-built condition, the increase in sample diameter fron 0.5 mm to 3 mm (Figure 2a–d) led to a coarser grained microstructure and a significantl stronger texture along the [001] direction parallel to the build direction (BD). The forma tion of strong textures and columnar grained microstructures is strongly promoted b the highest thermal gradient (G) and the solidification velocity (V) during processing as reported in numerous studies [21,22,26]. Thus, adjusting the process parameters allow for a microstructural design of metallic alloys, at least for a constant geometry and constan sample dimensions. However, in the case of fixed processing parameters, geometry affect G and V, at least when very small dimensions of cross-sections are considered.

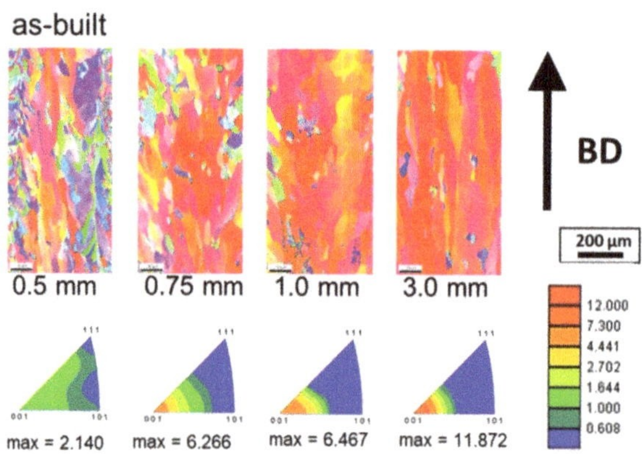

Figure 2. Electron backscatter diffraction (EBSD) inverse pole figure (IPF) maps of additively manu factured Fe-Mn-Al-Ni samples plotted in build direction (BD) for diameters ranging from 0.5 mm to 3 mm for the as-built condition (**a–d**). BD indicates the build direction. The color-coded triangles reveal the maximum texture. The relevant legend coding texture intensity is shown on the right side (max. 12.000).

From the results shown in Figure 2, it can be directly deduced that the variation in geometrical dimensions of the as-built parts is linked to a change in the evolution of the microstructure. As stated before, this is related to a change in G and V, both values being simultaneously affected by the volume energy, scanning strategy, and the build platform temperature. In an earlier study by some of the current authors, a build platform temperature of 200 °C was used during L-PBF processing, leading to a similar kind of microstructure transition, as compared to the current work [28]. However, in that study the strongest texture was observed in the smaller diameter cylinders and not, as shown in the present results, in samples featuring relatively large diameters. Since most L-PBF process parameters (laser power, scan velocity, etc.) were generally similar in both studies, the build platform temperature is supposed to have the most significant impact on the evolution of the microstructure at this point. It can be deduced that the higher build platform temperature led to a change in the overall thermal history during processing, directly affecting the solidification character, and eventually, the final grain morphology and orientation.

The applied heat treatment procedure employed to stimulate AGG is depicted in Figure 3a. Figure 3b–d shows the impact of the repetition of the HTC for 1, 3, and 5 times, for the given heat treatment procedure (Figure 3a). In Figure 3b, after 1 HTC, a columnar, fine-grained microstructure is seen. Already after 3 HTC (Figure 3c), a change towards a coarser grained microstructure is obvious. A large grain formed by consuming the initially columnar grains. After 5 HTC significant grain growth occurred, and a single grain of around 7 mm characterized by the <111> direction could be identified (Figure 3d). Thus, from the IPF maps presented, it was obvious that the AGG occurred in the additively manufactured Fe-Mn-Al-Ni. However, the results indicated certain inhomogeneities in the microstructure, i.e., the initial texture was not strong enough to be transferred into the abnormally grown grains.

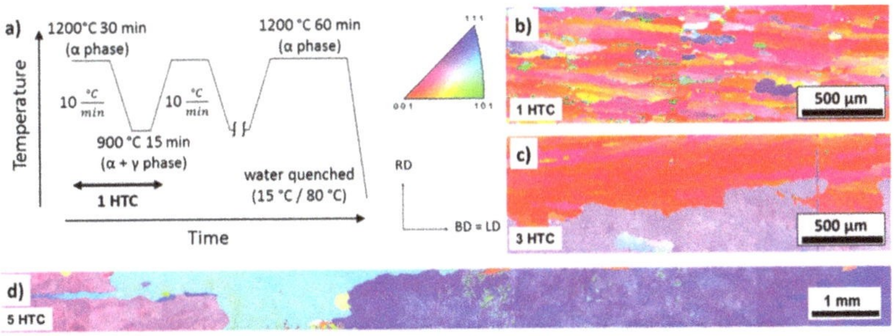

Figure 3. (**a**) Schematic highlighting a single heat treatment cycle (HTC) step (horizontal arrow) of the procedure employed to promote abnormal grain growth (AGG). (**b–d**) EBSD IPF maps depicting the cyclically heat-treated samples plotted in BD, after 1 HTC (**b**), 3 HTC (**c**), and 5 HTC (**d**).

Vickers hardness tests on all samples in the as-built condition, as well as after 1 and 5 HTC (without aging) were conducted (Figure 4). The as-built condition showed a noticeably higher hardness than those samples following 1 or 5 HTC. A similar result was observed in the already mentioned former study, where the hardness in the as-built condition increased up to 420 HV [28]. L-PBF processing at a build platform temperature of 200 °C resulted in the formation of nanoscaled β-phase precipitates in the L-PBF-processed Fe-Mn-Al-Ni alloy and, in consequence, led to a significant increase in hardness, as compared to a solution-treated condition. As detailed before, it was reported in several other studies (primarily focusing on conventionally processed material) that an artificial aging at 200 °C led to the formation of the β-phase precipitates, and thus, material strength and hardness can be substantially increased. In the present study, the elevated platform temperature of 500 °C obviously also had a significant impact on the hardness, since in the as-built condition, the hardness was even found to be higher than in the former study [28], i.e., 600 HV in present work (Figure 4). This difference clearly indicates that precipitates might also have formed during L-PBF processing at 500 °C. Moreover, it is likely that β-Mn phase evolved here. It was already reported for conventionally processed Fe-Mn-Al-Ni that above 300 °C, the β-Mn phase evolves in addition to the β-phase precipitates [33,34]. Taking into account that following L-PBF processing the samples were held at 400 °C for 15 min, it could be assumed that the high hardness values following L-PBF processing at 500 °C, were at least partly related to the formation of the β-Mn phase. However, the related increase in hardness did not seem to affect the formation of defects upon quenching, since no significant defect formation was observed in the material followin processing at 500 °C (Figure 1). In other words, the build platform temperature seeme to be high enough to impede any kind of process-related cracking mechanism during processing and subsequent cooling.

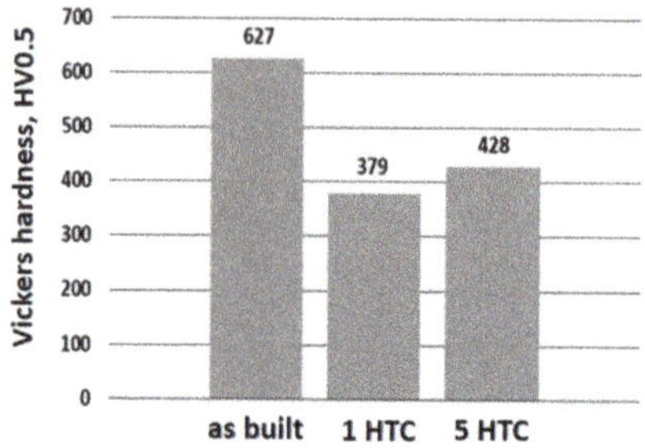

Figure 4. Vickers hardness of laser powder bed fusion (L-PBF) processed Fe-Mn-Al-Ni in the as-built condition, as well as after 1 and 5 HTCs.

In the course of any heat treatment routes considered, the as-built condition was solutionized and both the nano scaled β-precipitated and the β-Mn phase dissolved, eventually leading to a decrease in hardness to about 379 HV and 428 HV for 1 HTC and 5 HTC, respectively. This was in line with the findings presented in previous work [28]. Considering the grain growth induced by the HTC (Figures 2 and 3), the slight increase of hardness after 5 HTC (Figure 4) could be linked to the abnormally grown grains, i.e., the local effect of crystallographic grain orientation of abnormally grown grains, or marginal differences in quenching rates, eventually affecting the absolute value of Vickers hardness.

Analysis of the chemical composition of different conditions revealed a significant loss in the Mn content (Table 1) in the L-PBF processed condition. Whereas the nominal chemical composition contains 34 at. % of manganese, high-temperature L-PBF processing obviously results in a loss of around 3 at. % of Mn. Thus, a considerable change in the functional material properties, e.g., critical stresses for phase transformation, transformation temperatures, and phase stabilities, had to be expected. However, although this change in chemical composition is significant, the superelastic properties of the studied Fe-Mn-Al-Ni alloy are seen to be very promising (Figure 5).

Table 1. Nominal and experimentally determined chemical compositions of Fe-Mn-Al-Ni characterized in form of powder and L-PBF processed bulk material. The relative standard deviation (three measurements) was found to be below 1%.

Chemical Composition	Nominal	Powder	As Built
Ni (at. %)	7.50	7.49	7.87
Al (at. %)	15.00	15.44	15.88
Mn (at. %)	34.00	33.21	30.81
Fe (at. %)	43.50	42.52	44.86

Figure 5a depicts the stress–strain response obtained at RT under tensile loading for a Fe-Mn-Al-Ni sample initially processed by L-PBF. Such data are shown for the first time in literature. Following 5 HTC, good superelastic material properties can be seen (Figure 5), being similar to other stress–strain hysteresis curves shown in literature [35]. The critical stress for stress induced phase transformation (SIMT) was about 405 MPa. Following the first loading cycle up to 2% only about 0.2% irreversible strain could be identified. Following loading up to 4% about 0.6% residual strain accumulated. The relatively fast accumulation of residual strains indicates the activation of irreversible processes in parallel to phase transformation. Thus, the formation of dislocations capable of pinning the evolving martensite phase boundaries can be expected [36–38]. Analysis of such elementary mechanisms is currently under investigation and will be reported in follow-up studies. Up to now, it is assumed that the martensitic phase boundaries are mechanically stabilized and the back transformation into the austenite is suppressed, as was already revealed for conventionally processed material [36–38].

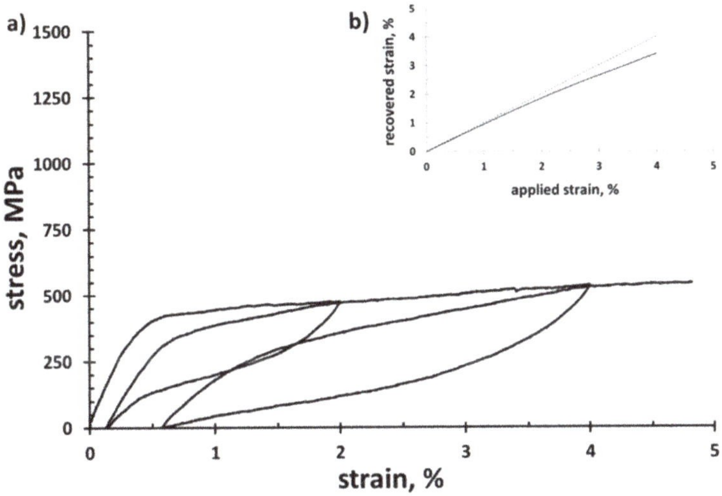

Figure 5. (**a**) Stress–strain response under tensile loading obtained from Fe-Mn-Al-Ni at RT after 5 HTC; (**b**) recovered strain plotted against the applied strain.

Figure 5b shows the applied strain plotted against the reversible strain. It can clearly be seen that the material shows a high reversibility upon loading in the very first superelastic cycle, i.e., loading to 2% strain. With further increase of the applied strain, the share of recovered strain decreases. This leads to the conclusion that the remaining grain constraints, imposed by the imperfect bamboo-like microstructure, dominated the evolution of residual strain. In this context, the loss of Mn is supposed to detrimentally affect the AGG process. Thus, the main challenge for further investigations will be the preservation of the chemical composition, as this can be seen as a key factor towards robust functional material properties in additively manufactured Fe-Mn-Al-Ni SMAs.

4. Conclusions

In the present work, laser powder bed fusion (L-PBF) processing of a Fe-Mn-Al-Ni shape memory alloy (SMA) at an elevated build platform temperature of 500 °C was conducted, followed by post-treatments, microstructure analysis, and tensile testing. The following conclusions could be drawn from the presented results:

1. L-PBF processing at 500 °C leads to crack-free Fe-Mn-Al-Ni bulk samples offering the possibility to fabricate individually designed, near-net-shaped parts by employing only one manufacturing step. This was not achieved so far due to the limited heating capabilities of most L-PBF systems available.
2. L-PBF processing leads to a noticeable change in chemical composition of the Fe-Mn-Al-Ni shape memory alloy. Mn is reduced by about 3 at. % upon L-PBF processing.
3. It is shown that a strong texture along the <001> direction primarily forms parallel to the build direction. However, the solidification microstructure is somehow affected by the geometry of the processed parts. In the as-built condition, an increase of the

diameter leads to a columnar grained microstructure and a strong texture in <001 direction.
4. Vicker's hardness measurements reveal a higher hardness in the as-built conditio This can likely be correlated to the formation of the β-Mn phase promoted by th elevated build platform temperature.
5. The application of a cyclic heat treatment promotes abnormal grain growth (AGG) i the L-PBF processed Fe-Mn-Al-Ni.
6. The microstructure established by the optimized heat treatment route is characterize by a good shape recovery ratio under tensile loading. However, elementary degr dation mechanisms still limit the full potential of the L-PBF processed Fe-Mn-Al-N A relatively fast accumulation of residual strains is seen promoted by prevailing grai constraints.

Author Contributions: F.C.E.: Visualization, Original Draft; F.B.: Conceptualization, Investigatio Review & Editing; T.G.: Conceptualization, Investigation, Review & Editing; M.V.: Conceptualizatio Review & Editing; P.K.: Conceptualization, Visualization, Review & Editing; T.N.: Conceptualizatio Review & Editing. All authors have read and agreed to the published version of the manuscript.

Funding: Financial support by DFG is acknowledged (Project No. 250216343).

Data Availability Statement: Data are available from the corresponding author on reasonable reques

Acknowledgments: T.G. thanks S. Pauly and U. Kühn for fruitful discussions, K. Schröder an N. Geißler for technical assistance, as well as A. Voß and H. Bußkamp for carrying out the chem ical analysis of the powder and the additively manufactured material. In addition, D. Becke and U. Biscop are thanked for their outstanding support regarding the development of the high temperature build envelope.

Conflicts of Interest: The authors declare no conflict of interest.

References

1. Mehrpouya, M.; Bidsorkhi, H.C. MEMS Applications of NiTi Based Shape Memory Alloys: A Review. *Micro Nanosyst.* **2017**, 8 79–91. [CrossRef]
2. Karaca, H.E.; Acar, E.; Tobe, H.; Saghaian, S.M. NiTiHf-based shape memory alloys. *Mater. Sci. Technol.* **2014**, *30*, 1530–1544. [CrossRef
3. Enkovaara, J.; Ayuela, A.; Zayak, A.T.; Entel, P.; Nordström, L.; Dube, M.; Jalkanen, J.; Impola, J.; Nieminen, R.M. Magneticall driven shape memory alloys. *Mater. Sci. Eng. A* **2004**, *378*, 52–60. [CrossRef]
4. Karaca, H.E.; Karaman, I.; Basaran, B.; Ren, Y.; Chumlyakov, Y.I.; Maier, H.J. Magnetic Field-Induced Phase Transformation in NiMnCoIn Magnetic Shape-Memory Alloys-A New Actuation Mechanism with Large Work Output. *Adv. Funct. Mater.* **2009**, *19* 983–998. [CrossRef]
5. Sato, A.; Chishima, E.; Soma, K.; Mori, T. Shape memory effect in γ⇄ε transformation in Fe-30Mn-1Si alloy single crystals *Acta Metall.* **1982**, *30*, 1177–1183. [CrossRef]
6. Tanaka, Y.; Himuro, Y.; Kainuma, R.; Sutou, Y.; Omori, T.; Ishida, K. Ferrous polycrystalline shape-memory alloy showing huge superelasticity. *Science* **2010**, *327*, 1488–1490. [CrossRef]
7. Omori, T.; Ando, K.; Okano, M.; Xu, X.; Tanaka, Y.; Ohnuma, I.; Kainuma, R.; Ishida, K. Superelastic effect in polycrystalline ferrous alloys. *Science* **2011**, *333*, 68–71. [CrossRef] [PubMed]
8. Sawaguchi, T.; Maruyama, T.; Otsuka, H.; Kushibe, A.; Inoue, Y.; Tsuzaki, K. Design Concept and Applications of Fe–Mn–Si-Based Alloys—From Shape-Memory to Seismic Response Control. *Mater. Trans.* **2016**, *57*, 283–293. [CrossRef]
9. Lee, D.; Omori, T.; Han, K.; Hayakawa, Y.; Kainuma, R. Effect of Thermomechanical Processing on Texture and Superelasticity in Fe–Ni-Co-Al-Ti-B Alloy. *Shap. Mem. Superelasticity* **2018**, *4*, 102–111. [CrossRef]
10. Lee, D.; Omori, T.; Kainuma, R. Ductility enhancement and superelasticity in Fe–Ni–Co–Al–Ti–B polycrystalline alloy. *J. Alloy Compd.* **2014**, *617*, 120–123. [CrossRef]
11. Omori, T.; Okano, M.; Kainuma, R. Effect of grain size on superelasticity in Fe-Mn-Al-Ni shape memory alloy wire. *APL Mater* **2013**, *1*, 32103. [CrossRef]
12. Ozcan, H.; Ma, J.; Wang, S.J.; Karaman, I.; Chumlyakov, Y.; Brown, J.; Noebe, R.D. Effects of cyclic heat treatment and aging on superelasticity in oligocrystalline Fe-Mn-Al-Ni shape memory alloy wires. *Scr. Mater.* **2017**, *134*, 66–70. [CrossRef]
13. Tseng, L.W.; Ma, J.; Wang, S.J.; Karaman, I.; Chumlyakov, Y.I. Effects of crystallographic orientation on the superelastic response of FeMnAlNi single crystals. *Scr. Mater.* **2016**, *116*, 147–151. [CrossRef]
14. Tseng, L.W.; Ma, J.; Wang, S.J.; Karaman, I.; Kaya, M.; Luo, Z.P.; Chumlyakov, Y.I. Superelastic response of a single crystalline FeMnAlNi shape memory alloy under tension and compression. *Acta Mater.* **2015**, *89*, 374–383. [CrossRef]

15. Vallejos, J.M.; Malarría, J.A. Growing Fe-Mn-Al-Ni single crystals by combining directional annealing and thermal cycling. *J. Mater. Process. Technol.* **2020**, *275*, 116317. [CrossRef]
16. Vollmer, M.; Arold, T.; Kriegel, M.J.; Klemm, V.; Degener, S.; Freudenberger, J.; Niendorf, T. Promoting abnormal grain growth in Fe-based shape memory alloys through compositional adjustments. *Nat. Commun.* **2019**, *10*, 2337. [CrossRef]
17. Vollmer, M.; Segel, C.; Krooß, P.; Günther, J.; Tseng, L.W.; Karaman, I.; Weidner, A.; Biermann, H.; Niendorf, T. On the effect of gamma phase formation on the pseudoelastic performance of polycrystalline Fe–Mn–Al–Ni shape memory alloys. *Scr. Mater.* **2015**, *108*, 23–26. [CrossRef]
18. Gibson, I.; Rosen, D.; Stucker, B. *Additive Manufacturing Technologies. 3D Printing, Rapid Prototyping and Direct Digital Manufacturing*; Springer: Berlin/Heidelberg, Germany, 2015.
19. Körner, C.; Helmer, H.; Bauereiß, A.; Singer, R.F. Tailoring the grain structure of IN718 during selective electron beam melting. In *MATEC Web of Conferences*; EDP Sciences: Les Ulis, France, 2014; Volume 14, p. 8001.
20. Thijs, L.; Kempen, K.; Kruth, J.-P.; van Humbeeck, J. Fine-structured aluminium products with controllable texture by selective laser melting of pre-alloyed AlSi10Mg powder. *Acta Mater.* **2013**, *61*, 1809–1819. [CrossRef]
21. Niendorf, T.; Brenne, F.; Schaper, M.; Riemer, A.; Leuders, S.; Reimche, W.; Schwarze, D.; Maier, H.J. Labelling additively manufactured parts by microstructural gradation—Advanced copy-proof design. *Rapid Prototyp. J.* **2016**, *22*, 630–635. [CrossRef]
22. Niendorf, T.; Leuders, S.; Riemer, A.; Richard, H.A.; Tröster, T.; Schwarze, D. Highly Anisotropic Steel Processed by Selective Laser Melting. *Met. Mater. Trans. B* **2013**, *44*, 794–796. [CrossRef]
23. Dadbakhsh, S.; Vrancken, B.; Kruth, J.-P.; Luyten, J.; van Humbeeck, J. Texture and anisotropy in selective laser melting of NiTi alloy. *Mater. Sci. Eng. A* **2016**, *650*, 225–232. [CrossRef]
24. Gustmann, T.; Neves, A.; Kühn, U.; Gargarella, P.; Kiminami, C.S.; Bolfarini, C.; Eckert, J.; Pauly, S. Influence of processing parameters on the fabrication of a Cu-Al-Ni-Mn shape-memory alloy by selective laser melting. *Addit. Manuf.* **2016**, *11*, 23–31. [CrossRef]
25. Haberland, C.; Elahinia, M.; Walker, J.M.; Meier, H.; Frenzel, J. On the development of high quality NiTi shape memory and pseudoelastic parts by additive manufacturing. *Smart Mater. Struct.* **2014**, *23*, 104002. [CrossRef]
26. Shiva, S.; Palani, I.A.; Mishra, S.K.; Paul, C.P.; Kukreja, L.M. Investigations on the influence of composition in the development of Ni–Ti shape memory alloy using laser based additive manufacturing. *Opt. Laser Technol.* **2015**, *69*, 44–51. [CrossRef]
27. Elahinia, M.; Shayesteh Moghaddam, N.; Taheri Andani, M.; Amerinatanzi, A.; Bimber, B.A.; Hamilton, R.F. Fabrication of NiTi through additive manufacturing: A review. *Prog. Mater. Sci.* **2016**, *83*, 630–663. [CrossRef]
28. Niendorf, T.; Brenne, F.; Krooß, P.; Vollmer, M.; Günther, J.; Schwarze, D.; Biermann, H. Microstructural Evolution and Functional Properties of Fe-Mn-Al-Ni Shape Memory Alloy Processed by Selective Laser Melting. *Met. Mat. Trans. A* **2016**, *47*, 2569–2573. [CrossRef]
29. Torrent, C.; Bauer, A.; Vollmer, M.; Niendorf, T. On the Challenges toward Realization of Functionally Graded Structures by Electron Beam Melting—Fe-Base Shape Memory Alloy and Stainless Steel. In *Structural Integrity of Additive Manufactured Materials and Parts*; Shamsaei, N., Seifi, M., Eds.; Selected technical papers STP 1631; ASTM International: West Conshohocken, PA, USA, 2020; pp. 20–33. [CrossRef]
30. Omori, T.; Iwaizako, H.; Kainuma, R. Abnormal grain growth induced by cyclic heat treatment in Fe-Mn-Al-Ni superelastic alloy. *Mater. Des.* **2016**, *101*, 263–269. [CrossRef]
31. Tseng, L.W.; Ma, J.; Hornbuckle, B.C.; Karaman, I.; Thompson, G.B.; Luo, Z.P.; Chumlyakov, Y.I. The effect of precipitates on the superelastic response of [100] oriented FeMnAlNi single crystals under compression. *Acta Mater.* **2015**, *97*, 234–244. [CrossRef]
32. Schwab, H.; Bönisch, M.; Giebeler, L.; Gustmann, T.; Eckert, J.; Kühn, U. Processing of Ti-5553 with improved mechanical properties via an in-situ heat treatment combining selective laser melting and substrate plate heating. *Mater. Des.* **2017**, *130*, 83–89. [CrossRef]
33. Frenck, J.-M.; Vollmer, M.; Mandel, M.; Krüger, L.; Niendorf, T. On the Influence of Microstructure on the Corrosion Behavior of Fe–Mn–Al–Ni Shape Memory Alloy in 5.0 wt% NaCl Solution. *Adv. Eng. Mater.* **2020**, 2000865. [CrossRef]
34. Huang, P.; Peng, H.; Wang, S.; Zhou, T.; Wen, Y. Relationship between martensitic reversibility and different nano-phases in a FeMnAlNi shape memory alloy. *Mater. Charact.* **2016**, *118*, 22–28. [CrossRef]
35. Tseng, L.W.; Ma, J.; Vollmer, M.; Krooß, P.; Niendorf, T.; Karaman, I. Effect of grain size on the superelastic response of a FeMnAlNi polycrystalline shape memory alloy. *Scr. Mater.* **2016**, *125*, 68–72. [CrossRef]
36. Gall, K.; Maier, H.J. Cyclic deformation mechanisms in precipitated NiTi shape memory alloys. *Acta Mater.* **2002**, *50*, 4643–4657. [CrossRef]
37. Vollmer, M.; Kriegel, M.J.; Krooß, P.; Martin, S.; Klemm, V.; Weidner, A.; Chumlyakov, Y.; Biermann, H.; Rafaja, D.; Niendorf, T. Cyclic Degradation Behavior of [001]-Oriented Fe–Mn–Al–Ni Single Crystals in Tension. *Shap. Mem. Superelasticity* **2017**, *3*, 335–346. [CrossRef]
38. Vollmer, M.; Krooß, P.; Kriegel, M.J.; Klemm, V.; Somsen, C.; Ozcan, H.; Karaman, I.; Weidner, A.; Rafaja, D.; Biermann, H.; et al. Cyclic degradation in bamboo-like Fe–Mn–Al–Ni shape memory alloys—The role of grain orientation. *Scr. Mater.* **2016**, *114*, 156–160. [CrossRef]

metals

MDPI

Article

Control of Density and Grain Structure of a Laser Powder Bed-Fused Superelastic Ti-18Zr-14Nb Alloy: Simulation-Driven Process Mapping

Vladimir Brailovski [1,*], Victoria Kalinicheva [2], Morgan Letenneur [1], Konstantin Lukashevich [2], Vadim Sheremetyev [2] and Sergey Prokoshkin [2]

1 Department of Mechanical Engineering, École de Technologie Supérieure, 1100 Notre-Dame Street West, Montreal, QC H3C 1K3, Canada; morgan.letenneur.1@ens.etsmtl.ca

2 Metal Forming Department, National University of Science and Technology MISIS, 119049 Moscow, Russia; m1902355@edu.misis.ru (V.K.); m142659@edu.misis.ru (K.L.); sheremetyev@misis.ru (V.S.); prokoshkin@tmo.misis.ru (S.P.)

* Correspondence: vladimir.brailovski@etsmtl.ca; Tel.: +1-514-396-8594

Received: 6 November 2020; Accepted: 14 December 2020; Published: 21 December 2020

Abstract: This study focuses on the control of density and grain structure of a superelastic Ti-18Zr-14Nb (at. %) alloy subjected to laser powder bed fusion. It starts with the production and characterization of a Ti-18Zr-14Nb powder feedstock and printing of a series of calibration specimens. These specimens are next subjected to chemical, structural, phase and texture analyses in order to collect experimental data needed to build simulation-driven processing maps in the laser energy density–material build rate coordinates. The results of this study prove that, once calibrated, the simulation-driven processing maps can be used to relate the main LPBF parameters (laser power, scanning speed, hatching distance and layer thickness) to the density and grain structure of the printed material, and the process productivity (build rate). Even though this demonstration is made for a specific material–system combination (TiNbZr & TruPrint 1000), such a process mapping is feasible for any material–system combination and can, therefore, be exploited for the process optimization purposes and for manufacturing of functionally graded materials or parts with intentionally seeded porosity.

Keywords: additive manufacturing; laser powder bed fusion; shape memory alloys; density control; structure control; process simulation

1. Introduction

Over the past decade, additive manufacturing came out on top when selecting a manufacturing technique capable of producing metallic parts with complex geometry [1]. This technology allows an easy scaling from a prototype to a mass-customized product, which is frequently needed in the field of biomedical implants [1–3]. Additive manufacturing is especially suited for the manufacture of porous (cellular or lattice) structures, which offer a unique possibility of expanding the effective range of functional properties attainable with convectional materials. Several manufacturing techniques could be used to produce lattice structure with controlled pore and strut/wall size, shape, orientation and distribution, including the creation of hierarchical and functionally graded lattice structures [4]. Among them, laser powder bed fusion (LPBF) offers the best-in-class control of geometric attributes and mechanical properties of such structures made of a multitude of metal and metallic alloys. In this process, a high-power laser follows a certain path, determined by a selected laser scanning strategy and a numerical model of the part, to consolidate the part by fusing powder particles layer-by-layer [5].

Shape memory alloys (SMAs) provide additional functionality to common materials, such as the superelasticity and shape memory effects. For example, in medicine, porous superelastic implants can reduce the risk of complications associated with their integration into the implantation site [6], and such materials can be 3D printed and morph over time under the influence of thermal, mechanical or other external stimuli (4D printing [7]). In terms of their functional characteristics, titanium nickelide (Ti-Ni) SMAs are unsurpassed (recoverable strains of up to 10% and recovery stress of up to 1200 MPa), but they contain ~50% of nickel, which makes the release of this toxic metal into the human body a possibility [8].

In this context, the best candidates for long-term biomedical applications are metastable β nickel-free Ti alloys, such as Ti-Nb-(Zr, Ta or Mo) [9–12]. They can manifest either temperature- or stress-induced reversible thermoelastic $\beta \leftrightarrow \alpha''$ transformations, with the former being responsible for the shape memory effect, while the latter mimicking a low-stiffness plateau-like superelastic behavior of biological tissues [8,9]. The results of a background research confirm the possibility of producing these alloys with relatively high recovery stresses (600–700 MPa) and recoverable strains (6–7%), but without the toxic nickel [12].

Among numerous metastable Ti alloys, Ti-18Zr-14Nb (in at. %) represents one of the best candidates for biomedical implants because it fulfills the following four criteria: (a) biocompatibility of all constitutive elements, (b) recoverable strain at body temperature (superelasticity) comparable to that of the reference Ti-Ni SMA, (c) mechanical properties comparable to those of the commonly used $\alpha + \beta$ Ti-6Al-4V alloy and (d) Young's modulus close to that of cortical bone [9].

It appears evident that new opportunities for the development of high-performance biomedical implants could be envisaged by synergistically combining the exceptional functional properties of superelastic nickel-free Ti alloys with the design flexibility offered by additive manufacturing. It should, however, be noted that the functional properties of SMAs (temperature range of martensitic transformation, recoverable strains and recovery stresses) are much more structural and, therefore, processing-dependent than the mechanical properties of common materials [13,14]. For example, increasing laser energy density of the LPBF process may result not only in higher impurity content but also in the evaporation of certain alloy elements [7], which makes the control of the SMA transformation temperature range difficult. Another difficulty related to the use of LPBF arises from processing-induced porosity. The latter is frequently due to suboptimal technological conditions of laser fusion and a limited quality of raw materials (presence of inclusions and inadequate powder particle size distribution and morphology). The problem of residual porosity can be attenuated by subsequent hot isostatic pressing, but this approach is not without drawbacks, such as undesirable structure coarsening, for example [15]. Other processing-induced flaws, such as cracks, microstructural heterogeneity and segregation, could also impact the service properties of printed components and must, therefore, be thoroughly controlled [16].

It is also known that the as-built density and microstructure of printed components may be controlled by selecting a right combination of LPBF processing parameters, such as laser power, scanning speed, hatching distance and layer thickness. To produce components with a controlled density, grain structure, phase state and texture, the processing–structure interaction must be studied for each new powdered material, and this task is almost impossible to complete without numerical simulations [17,18]. In this context, a simplified numerical algorithm allowing rapid melt pool simulations for thousands of printing conditions was developed and experimentally validated for Fe, AlSiMg, IN625, Ti64 and 316L alloys. It was then successfully applied to control the processing-induced porosity [19–21] and the grain structure [22] of the selected alloys.

The above combination of the melt pool modeling and the design of experiment approach has also been applied to the Ti-18Zr-14Nb superelastic alloy, but without a specific focus on the grain structure of printed components [23]. To begin filling this gap, this study starts with the characterization of a Ti-18Zr-14Nb powder feedstock. Next, a processing window for the selected LPBF system is defined, and calibration specimens are printed using different sets of laser powder fusion parameters, strategically distributed within the limits of the processing window. Furthermore, the printed specimens are subjected to chemical, structure, phase and texture analyses. Finally, all the experimental data obtained are employed to build the density and the grain structure processing maps using a proprietary simulation-driven algorithm.

2. Methodology

2.1. LPBF Process Calibration Experiment

In this study, a TruPrint 1000 (TRUMF GmbH, Ditzingen, Germany) laser powder bed fusion system equipped with a 200 W ytterbium fiber laser (spot diameter of 30 μm) was used. As a first step, the processing window was delimited by varying the main processing parameters of this printing system within their operational borders: the laser power from 40 to 200 W, the laser speed from 200 to 4000 mm/s and the hatching space from 60 to 200 μm, while keeping a constant layer thickness of 30 μm. This processing window was built using the energy density (*E*)–build rate (*BR*) coordinates, as shown in Figure 1:

$$E\left(J/mm^3\right) = \frac{P}{v \times h \times t};\ BR\left(cm^3/h\right) = v \times h \times t \tag{1}$$

where *p* (W) is the laser power, *v* (m/s) is the laser speed, *h* (μm) is the hatching space and *t* (μm) is the layer thickness.

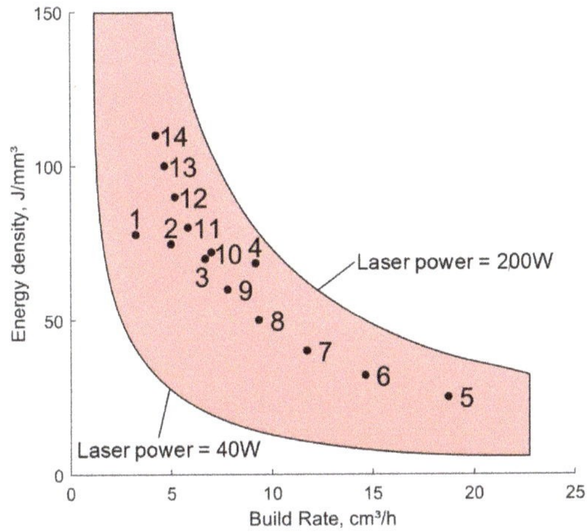

Figure 1. Energy density–build rate window with processing conditions of 14 Ti-18Zr-14Nb calibration specimens (*TruPrint 1000*).

Next, within the limits of this E–BR window, the processing conditions for fourteen (14) calibration specimens were determined (see Table 1). These specimens were regrouped into two sets. The main idea behind the use of the two sets of specimens was to cover the entire process window as widely as possible with the smallest number of specimens to be printed and analyzed. With this objective in mind, the first set, Specimens 1 to 4, was designed to first guess the conditions leading to the printing of a fully dense Ti-18Zr-14Nb alloy. This set was established by keeping the laser energy density in the process midrange of 73 ± 5 J/mm^3, while varying the build rate from 3.2 to 9.2 cm^3/h. The second specimen set, Specimens 5 to 14, was designed to explore the influence of LPBF processing conditions on the main subjects of this study: the density and the grain structure of printed material. This set was established by varying both the laser energy density (25–110 J/mm^3) and the build rate (18.7–4.3 cm^3/h) as widely as possible, while keeping the laser power constant (130 W).

Table 1. Processing parameters for 14 Ti-Zr-Nb calibration specimens (constant layer thickness of t = 30 μm).

Set	Specimens	Power, W	Speed, m/s	Hatch Distance, mm	E, J/mm^3	BR, cm^3/h
1st set	1	70	500	0.06	77.8	3.2
	2	104	772	0.06	74.7	5.0
	3	140	1080	0.06	72.0	7.0
	4	175	1420	0.06	68.5	9.2
2nd set	5	130	1576	0.11	25.0	18.7
	6	130	1231	0.11	32.0	14.6
	7	130	985	0.11	40.0	11.7
	8	130	788	0.11	50.0	9.4
	9	130	657	0.11	60.0	7.8
	10	130	563	0.11	70.0	6.7
	11	130	492	0.11	80.1	5.8
	12	130	438	0.11	89.9	5.2
	13	130	394	0.11	100.0	4.7
	14	130	358	0.11	110.0	4.3

2.2. Powder Production and Analysis

To obtain the powder feedstock used in this study, a 50 mm-diameter, 600 mm-long Ti-18Zr-14Nb alloy ingot supplied by Flowserve Corp. Comp (Irving, TX, USA) was gas-atomized by TLS Technik GmbH (Bitterfeld-Wolfen, Germany). The particle size distribution was measured using a Shimadzu SALD 7500 nano (Shimadzu Corp., Kyoto, Japan) laser scattering particle size analyzer. The powder morphology was qualified using a TESCAN VEGA LMH scanning electron microscope (TESCAN, Brno, Czech Republic).

2.3. LPBF Printing and Analyses of Printed Specimens

Fourteen 10 × 10 × 10 mm cubic specimens were printed and contoured (Figure 2a,c). The entire print field was divided into 4 × 4 mm squares. Each square was successively filled with tracks in 60/110 μm increments. To reduce distortions, a 90° hatch rotation scanning strategy between two successively built layers was adopted (Figure 2b).

After printing, the as-built cubic specimens were cut from the platform and subjected to the following analyses: chemical composition, density, phases, grain structure and crystallographic texture. The chemical composition of major elements in the powder and the printed specimens was measured using a TESCAN VEGA LMH scanning electron microscope (TESCAN, Brno, Czech Republic) equipped with an energy-dispersive X-ray microanalyzer (XCITE, Oxford Instruments, Abingdon-on-Thames, Royaume-Uni). The oxygen and nitrogen contents were controlled using a LECO TC600 analyzer. The density of the printed specimens was measured using Archimedes's technique (ASTM B962) and presented in percentage to the Ti-18Zr-14Nb alloy bulk density (5.66 g/mm^3).

Samples for the phase and structural analyses were cut from the printed specimens. The cross-sectional surfaces of the samples were prepared using an ATM Saphir 560 grinding/polishing machine (ATM GmbH, Mammelzen, Germany). Grinding (2 min, 10 N) was carried out on SiC paper with a grain size of 320–600–1200–2400 mesh. Polishing (20 min, 30 N) was realized using Eposil F suspensions (silicon oxide with a particle size of 0.1 μm) deposited on an ATM Iota cloth. During the polishing process, ammonia, hydrogen peroxide, liquid soap and diamond lubricant were added to improve the surface quality. Samples for the X-ray diffractometry analysis (XRD) and optical and scanning electron microscopy were polished mechanically and then etched in $1HF:3HNO_3:6H_2O$ solution to remove a damaged surface layer.

To study structural defects in the printed specimens, a Union Versamet-2 optical microscope equipped with a Nikon D90 camera (Nikon, Tokyo, Japan) was used, whereas an Ultima IV Rigaku diffractometer (monochromatic CuKα-radiation; 2θ ranges from 30 to 90 deg) was employed to study their phase state. To study a grain structure and texture, a TESCAN VEGA LMH scanning electron microscope (SEM) equipped with an electron backscatter diffraction (EBSD) unit was used. The texture evolution was characterized using Oxford Instruments' AZtec EBSD software (Oxford Instruments, Abingdon-on-Thames, Royaume-Uni) by tilting the samples by 70° and scanning them at 20 kV with a 2 μm step. The average grain size was measured from the EBSD images using a standard linear intercept procedure [24]. The measurements were performed on the vertical (parallel to the build direction, d_{xz}) and horizontal (perpendicular to the build direction, d_{xy}) faces of the specimens, and the grain aspect ratio GAR = d_{xz}/d_{xy} was calculated.

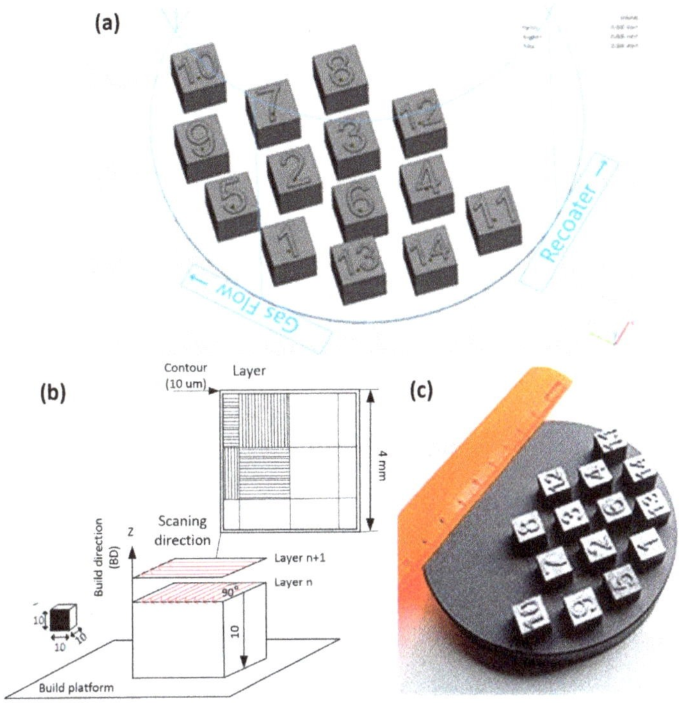

Figure 2. (**a**) Schematic presentation of the specimens on the build plate, (**b**) LPBF scanning strategy (dimensions in mm) and (**c**) result of the build.

2.4. Printed Density and Grain Structure Processing Maps

The experimental results relating to the density and grain structure of printed specimens were used to build the density, the grain size and the grain aspect ratio processing maps for a TruPrint 1000 LPBF system. To this end, a combined melt pool model and experimental design was applied (see details in [20,22]). In this approach, LPBF melt pool temperature field calculations were carried out using the analytical model (École de technologie supérieure, Montreal, Canada) of a semi-infinite solid with a moving Gaussian heat source [25]. This model has been successfully used for the density prediction in specimens manufactured from FeC [19], Ti-Zr-Nb [21], AlSiMg, IN625, Ti64 and 316L alloy powders [20]. The Gaussian model involves a symmetrical distribution of laser irradiance across the beam. The energy from the laser is assumed to be applied on the powder bed surface for a time interval defined by the scanning speed and the laser spot size. The physical properties of a Ti-Zr-Nb alloy in the bulk and powdered form used to calculate the temperature distribution $T(x, y, z)$ in the powder bed are collected in Table 2. The powder bed density (φ) was estimated based on the measurements reported in [23]. Finally, to establish the relationship between the LPBF processing parameters and the printed material density and grain structure, the correlation equations developed in [20,22] were used.

Table 2. Physical properties of the Ti-Zr-Nb alloy in bulk and powdered forms used for process modeling.

Material Characteristics	Bulk	Powder (t = 30 μm; φ = 60%)
Density, kg/m^3	5660	3396
Thermal conductivity *, W/mK	7.2	3.6
Specific heat *, J/kg.K	560	560
Laser absorptivity **	0.46	0.91
Melting temperature ***, °C	1650	1650

* Values correspond to Ti-6Al-4V (AZoMaterials, 2002); ** Calculated with equations provided in Letenneur et al. (2020); *** Calculated in Prokoshkin et al. (2016).

3. Results and Discussion

3.1. Powder Analysis

The particle size distribution (PSD) of the gas-atomized Ti-18Zr-14Nb powder is shown in Table 3 (d_{10} = 11 μm, d_{50} = 26 μm, d_{90} = 51 μm); the apparent powder density is 2.8 g/cm^3 (ISO 3923-1:2008). The majority of the particles are under 25 μm in equivalent diameter and a large number of small particles are attached to large particles, thus forming satellites (Figure 3a). A fine solidification microstructure is observed in powder particles, and larger particles demonstrate a cellular microstructure with an average cell size of ~1.5 μm (Figure 3b).

Table 3. Chemical compositions of the Ti-Zr-Nb powder and selected printed specimens.

Printing Conditions		Chemical Composition				
		Main Elements (at. %)			Interstitial Elements (wt. %)	
Specimen	Power Density (W/mm^3)	Ti	Zr	Nb	O	N
3	72	65.5	18.9	15.6	0.213	0.011
4	68.5	65.6	18.8	15.5	0.218	0.019
5	25	65.6	19.0	15.4	0.200	0.010
14	110	64.7	19.5	15.8	0.213	0.025
Powder		68.6	17.8	13.6	0.189	0.013

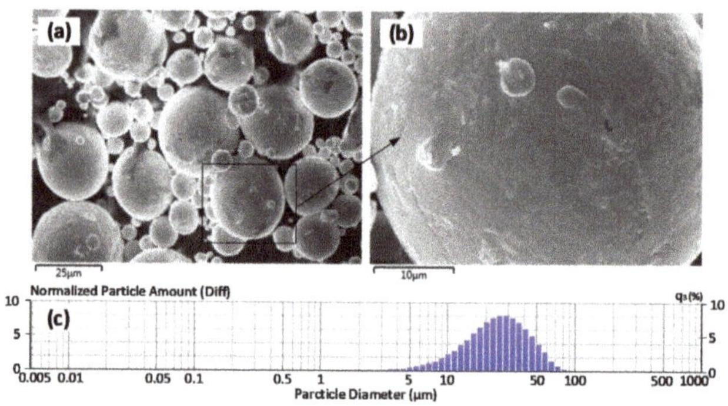

Figure 3. SEM images of gas-atomized Ti-18Zr-14Nb alloy powder: (**a**,**b**) morphology and microstructure of powder particles; (**c**) particle size distribution (PSD) of the Ti-18Zr-14Nb powder.

3.2. Analyses of Printed Specimens

3.2.1. Chemical Composition

The concentrations of the main and interstitial elements in the Ti-Zr-Nb powder and selected printed specimens (3, 4, 5, 14) are shown in Table 3. It can be seen that the printed specimens show lower titanium contents than the precursor powder, while the oxygen and nitrogen concentrations reveal the opposite trend. In Specimen 14, which is printed with the highest energy density of this study (E = 110 J/mm^3), both phenomena are the most striking: almost a 3.0 at. % loss in titanium is accompanied by 0.025 and 0.012 (wt. %) gains in oxygen and nitrogen, respectively. When printing with higher energy densities, this combined effect could lead to a significant depression of the martensite start temperature.

3.2.2. Density

Figure 4 shows the relative density of the printed specimens as a function of the laser energy density. It can be observed that the laser energy density must be ranged between 40 and 70 J/mm^3 to obtain a material density >99.5%. At lower and higher energy densities, the level of porosity increases (the numerical values of printed density are reported in Section 3.2.5). It should be noted that pores formed during low-energy printing are called the lack-of-fusion pores. These pores are filed with unmelted powder particles, generally follow the laser scan tracks and are more critical from a mechanical resistance viewpoint, as compared to uniformly distributed high energy-printed pores filled with gas [26]. Note also that the 40–70 J/mm^3 energy density range corresponds to many specimens of this study, thus primarily reflecting the robustness of the LPBF process to produce dense components. Secondly, this indicates the opportunity for the optimization of the process not only in terms of printed density, but also considering other criteria, such as material isotropy and process productivity, for example, as it will be shown later.

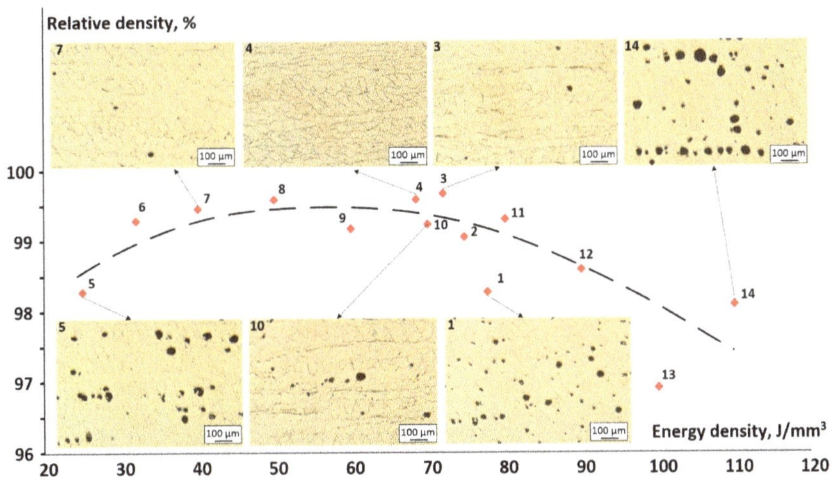

Figure 4. Relative density of the Ti-18Zr-14Nb specimens as a function of the laser energy density.

3.2.3. Phases

X-ray diffractograms of five selected specimens printed with different sets of processing parameters (4, 5, 7, 10 and 14) are shown in Figure 5. These specimens are separated into two groups with the first group (Specimens 4 and 10) printed with close levels of energy density, but different build rates, and the second group (Specimens 5, 7 and 14) printed with gradually increasing energy densities. An analysis of the diffraction patterns shows that the main phase component in all the specimens is β-phase; no additional peaks that could be attributed to any other phase are observed. The lattice parameters of β-phase calculated using the extrapolation method [27] do not significantly differ from each other (Table 4). Their average value of 0.3348 ± 0.0003 nm is, however, somewhat larger than that for the same alloy subjected to conventional processing, 0.3343 ± 0.0002 nm [28]. This discrepancy appears to be a consequence of the Zr- and Nb-enrichment of solid solution during processing (see Table 3).

The second explanation is a high level of oriented residual stresses of the dislocation substructure generated during solidification of printed specimens [29].

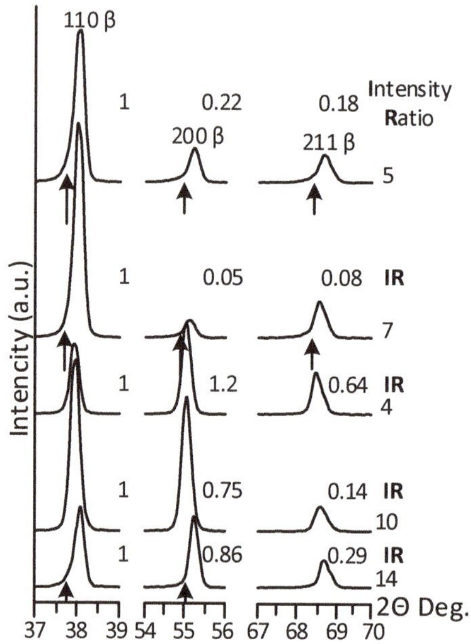

Figure 5. X-ray diffractograms of the selected as-built Ti-18Zr-14Nb samples; increasing power density from top to bottom.

Table 4. Beta-phase lattice parameters and X-ray peak intensity ratios of the selected as-built samples presented in decreasing power density order.

Specimen	E, J/mm^3	BR, cm^3/h	*a*, nm	Peak Intensity Ratio I_{110}:I_{200}:I_{211}
14	110	4.3	0.3347 ± 0.0003	1:0.88:0.42
10	70	6.7	0.3347 ± 0.0001	1:0.88:0.22
4	68.5	9.2	0.3351 ± 0.0001	1:1.32:0.85
7	40	11.7	0.3349 ± 0.0001	1:0.10:0.27
5	25	18.7	0.3347 ± 0.0003	1:0.27:0.26

The peak intensity ratio of Specimen 7 is typical for a state close to isotropic: very strong {110}, weak {200} and medium-strong {211} (Figure 5 and Table 4). The peak intensity ratios in Specimens 4, 10, 14 are typical for a state with a strong texture (in this case, the {100} plane is oriented mainly parallel to the irradiated surface, producing abnormally weak {110} and abnormally strong {200} peaks), or/and large grains. There are also signs of solid solution inhomogeneity, which is manifested by a slight extension of the left side of the peak soles towards smaller 2θ angles (indicated by arrows in Figure 5). Distortions of the X-ray lines are probably caused by the existence of Zr- and or Nb-enriched zones (solid solution separation phenomenon), which are more pronounced in Specimens 5 and 7, printed with lower energy densities and higher build rates [23].

3.2.4. Microstructure and Crystallographic Texture

EBSD analysis is performed for the same five specimens as those used for the XRD analysis: 4, 5, 7, 10 and 14. The microstructure of the selected samples in the E–BR coordinates is presented in Figure 6, where black lines correspond to high angle boundaries, i.e., angles >15°. The results of the grain structure analysis, including the grain size in the vertical and horizontal planes, the grain aspect ratio and the maximum texture intensity, are presented in decreasing power density order in Table 5. In can be seen that, overall, the lower the energy density and the greater the build rate, the finer and the more equiaxed the microstructure (see Figure 6 and Table 6).

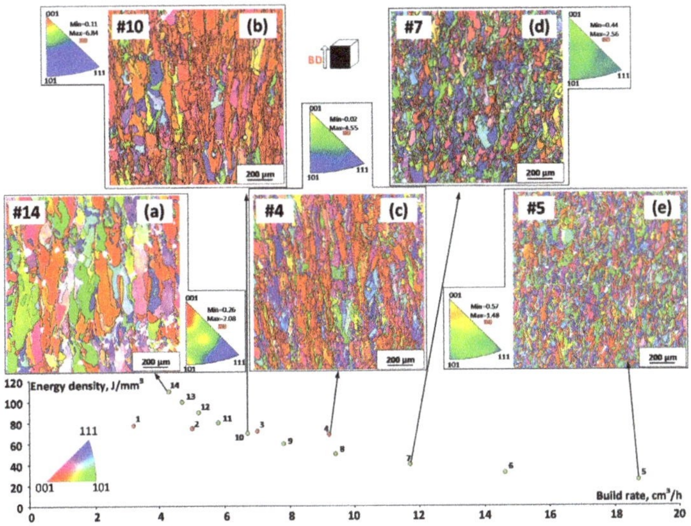

Figure 6. Microstructure and inverse pole figures of the as-built alloy presented in decreasing laser power density order: (**a**) 14, (**b**) 4, (**c**) 10, (**d**) 7 and (**e**) 5.

Table 5. Results of the structural analysis of selected specimens presented in decreasing laser power density order.

Specimen	E, J/mm^3	BR, cm^3/h	d_{xz}, µm	d_{xy}, µm	GAR	Text Int (Max)
14	110	4.3	199 ± 6	50 ± 6	2.88	2.08
10	70	6.7	151 ± 7	41 ± 5	3.28	6.84
4	68.5	9.2	97 ± 7	37 ± 4	3.17	4.55
7	40	11.7	66 ± 3	30 ± 4	2.62	2.56
5	25	18.7	50 ± 3	31 ± 4	2.34	1.48

Table 6. Measured and calculated density, grain size and grain aspect ratio, and their deviations.

Specimen	E, J/mm^3	BR, cm^3/h	Density, %			Grain Size (XZ), µm			Grain Aspect Ratio		
			Meas.	Calc.	Dev.	Meas.	Calc.	Dev., %	Meas.	Calc.	Dev., %
1	77.8	3.2	98.29	99.64	1.35	-	75	-	-	2.86	-
2	74.7	5.0	99.07	99.64	0.57	-	76	-	-	2.88	-
3	72.0	7.0	99.69	99.64	0.05	-	85	-	-	3.01	-
4	68.5	9.2	99.61	99.64	0.03	97 ± 7	97	0	3.17	3.19	0.6
5	25.0	18.7	98.30	98.30	0.00	50 ± 3	54	7.4	2.34	2.28	2.6
6	32.0	14.6	99.31	98.93	0.39	-	65	-	-	2.45	-
7	40.0	11.7	99.48	99.41	0.07	66 ± 3	78	15.4	2.62	2.65	1.1
8	50.0	9.4	99.61	99.61	0.00	-	94	-	-	2.86	-
9	60.0	7.8	99.19	99.58	0.39	-	112	-	-	3.06	-
10	70.0	6.7	99.25	99.29	0.05	151 ± 7	129	14.6	3.28	3.25	2.5
11	80.1	5.8	99.32	99.29	0.03	-	147	-	-	3.43	-
12	89.9	5.2	98.61	98.73	0.12	-	166	-	–	3.60	-
13	100.0	4.7	96.93	98.73	1.83	-	186	-	-	3.77	-
14	110.0	4.3	98.11	98.03	0.08	199 ± 6	X	X	2.88	X	X

X: bimodal grains distribution, the model is nonapplicable.

Overall, the higher the laser energy density, the coarser and the more elongated the grain structure; conversely, the lower the laser energy density, the finer and the more equiaxed the grain structure. However, the grain structures of Specimens 10 and 4, printed with similar energy densities (E = 70 and 68.5 J/mm^3, respectively), but different build rates (BR = 6.7 and 9.2 cm^3/h, respectively), are not identical: Specimen 4, printed with a higher build rate, shows a significantly finer grain structure than Specimen 10, built with a lower build rate: $d_{xz} = 97 \pm 7$ μm for the first, as compared to $d_{xz} = 151 \pm 7$ μm, for the second. These observations are supported by similar ones found in [30] and are explained by the fact that for a given laser energy density, the greater the build rate, the smaller the melt pool, the higher the cooling rate, and therefore, the finer the microstructure. The inverse pole figures of the as-built specimens are also presented in Figure 6. Overall, the lower the energy density, the higher the build rate, and the less textured the material, which is manifested by a decreasing <100>||BD direction texture intensity from 6.84, for Specimen 10, to 1.48, for Specimen 5.

An apparent exception is Specimen 14, printed with the highest energy density and the lowest build rate and containing the highest level of porosity among all the specimens of this study, ~2%. This specimen manifests a moderately elongated grain aspect ratio of 2.88 and a relatively low texture intensity of 2.08. The reason for this discrepancy for Specimen 14 stems from a significant amount of small equiaxed grains surrounding pores and shifting the average grain aspect ratio to lower values. A similar bimodal grain size distribution in highly porous printed specimens was also observed in [31].

3.2.5. Porosity, Grain Size and Grain Aspect Ratio Processing Maps

Exploiting the obtained experimental results, the density and grain structure processing maps are built using a simulation-driven approach described in [19,21,22] (see Figure 7). These maps demonstrate that the laser energy density and the build rate must range from 50 to 75 J/mm^3 and from 3 to 10 cm^3/h (Figure 7a), respectively, to print parts with the highest possible material density. Moreover, both the grain size and the grain aspect ratio decrease when the build rate increases and the laser energy density decreases (Figure 7b,c). The discrepancies between the experimentally measured and the numerically predicted densities, grain sizes and grain aspect ratios are calculated and collected in Table 6, and they do not exceed 2% for the printed density, 15% for the grain size and 3% for the grain aspect ratio (excluding Specimen 14). As a matter of fact, Specimen 14 manifests a bimodal grain size distribution as mentioned previously, which cannot be predicted by the model developed in the framework of this study.

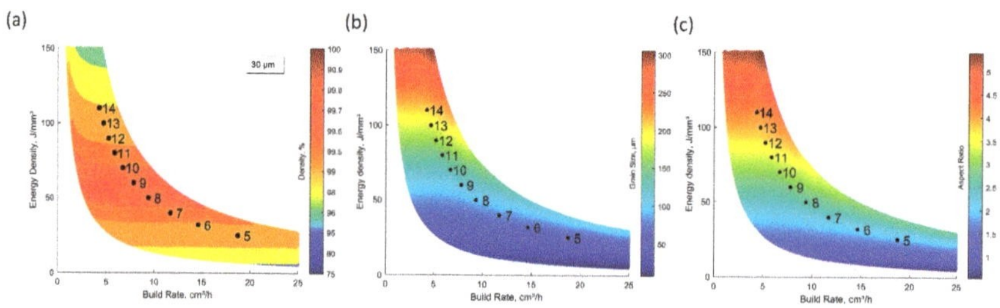

Figure 7. Processing maps for the Ti-Zr-Nb alloy: (**a**) density, (**b**) grain size and (**c**) grain aspect ratio; (hatching distance 110 μm; Specimen 14 is shown in (**a**), but not in (**b**,**c**) (explanations in text)).

Maps presented in Figure 7 can therefore be used to control the grain size and the grain aspect ratio of specimens printing on TruPrint 1000 LPBF systems from Ti-Zr-Nb powders. It is important to note that in order to simplify the representation, the maps shown in Figure 7 have only been plotted for a hatching space of 110 μm, which corresponds to Specimens 5 to 14 (the position of Specimen 14 is shown on the density map only, see previous explanations). The same methodology can be used to plot processing maps for other values of hatching space.

4. Discussion

The processing maps built using the experimental results of this study allow the selection of LPBF processing parameters with different objectives in mind. For example, the capacity of printing components with a controlled level of porosity could be used to build a database containing information on how the processing-induced porosity impacts the mechanical properties of printed components. This database could in turn serve for the quality control of printed components [31]. Moreover, the capacity to control the grain size and the grain aspect ratio could be used to obtain an appropriate balance between the structure features of printed components and the process productivity. Note, however, that these processing maps are only valid for the powder and the LPBF system of this study. If the same powder must be used with another LPBF system, these maps must be adjusted by printing a series of calibration specimens similar to those printed in this study. The reasons for this could be divided into two groups: the first, considering the powder spreading conditions, and the second, considering the laser printing conditions. Since each LPBF system is equipped with a specific powder spreading mechanism (roller, blade, etc.), the powder bed density and uniformity would differ, which in turn would inevitably influence the amount of laser energy absorbed by the powder and, therefore, the printing outcome. Moreover, the laser exposure and heat exchange conditions would also be different and dependent on the laser power and laser focus diameter, the build chamber geometry, the building plate temperature, the protective gas flow conditions, etc. Globally, more than 130 parameters would influence the printing outcome [32].

To support this point, consider the same Ti-Zr-Nb powder but two different LPBF systems: a TruPrint 1000 of this study and an EOS M280 from our preceding work [21]. The grain size and grain aspect ratio processing maps for these systems are compared in Figure 8a,c and Figure 8b,d, respectively. It can be seen that the higher the maximum laser power of a system, the larger the available processing window. Next, when comparing the grain structures of specimens printed with these two systems using similar printing conditions (TruPrint 1000 with E = 68.5 J/mm^3 and BR = 9.2 cm^3/h, and M280 with E = 60 J/mm^3 and BR = 12 cm^3/h), the first printing results in a grain structure that is significantly finer, while more elongated, than the second printing can be observed: TruPrint 1000 with $d_{xz} \approx 100$ µm and GAR ≈ 3.3 as compared to M280 with $d_{xz} \approx 200$ µm and GAR ≈ 2.5 (Figure 8a–d). These discrepancies could, at least partially, be explained by the differences in the laser exposure conditions: a smaller laser focus diameter (∅50 µm) in the first system as compared to a larger laser focus diameter (∅100 µm) in the second system results in smaller melt pools, higher cooling rates and finer grain structures. This difference is definitely not the only one that impacts the printing outcome, but since it is virtually impossible to numerically simulate all the discrepancies between different LPBF systems, experimental calibration of a specific system represents the only viable solution. However, once calibrated, simulation-driven processing maps of this study can be generated for any material–system combination and can be exploited for the process optimization purposes and can also be used to manufacture functionally graded materials [33] or parts with intentionally seeded porosity [15,34].

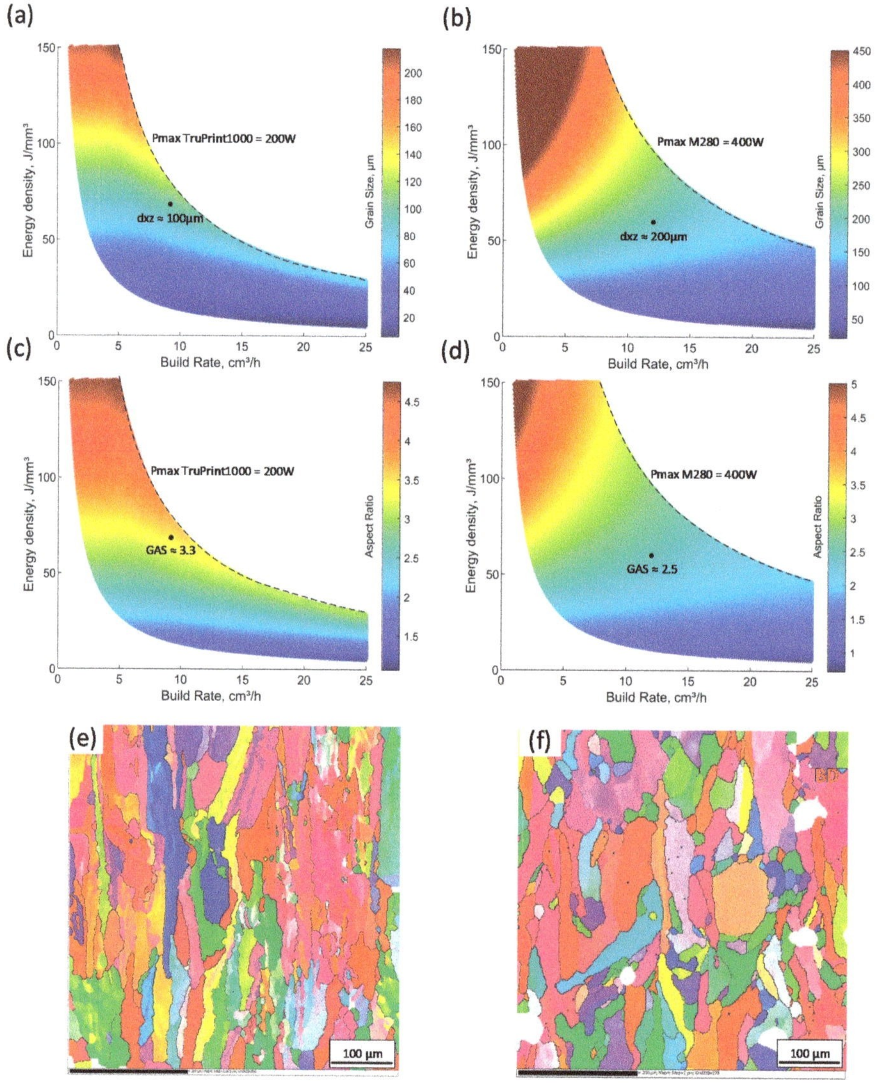

Figure 8. Microstructure and inverse pole figures of the Ti-Zr-Nb alloy printed using different LPBF systems: (**a**,**c**,**e**) TruPrint 1000 (Trumpf) and (**b**,**d**,**f**) M280 (EOS).

Finally, the recent development of AI looks to be a promising solution for many modeling issues faced in additive manufacturing. Although the use of this approach is currently limited to in-process tuning of the processing parameters [35], it could eventually be able to solve such multiprinter/multipowder modeling issues.

It must also be mentioned that the as-printed Ti-Zr-Nb alloy cannot be used without postprocessing treatment. In addition to a stress relief thermal treatment commonly applied to all LPBF parts to prevent their distortions during removal from the build plate, structural conditions promoting thermoelastic phase transformation must be created by subjecting printed specimens to a specific thermal treatment. The latter is necessary to allow benefiting from the functional properties of shape memory and superelasticity of these materials. It is known that for conventional mechanically deformed metastable near-beta Ti-based alloys, this thermal treatment involves 30–60 min annealing in the 450–550 °C temperature range, followed by rapid cooling [35]. However, since the thermal histories of deformed alloys and their laser powder bed-fused equivalents differ, the influence of postprocessing thermal treatments on the structure and properties of printed specimens must be studied and optimized accordingly. This last issue is actually under investigation.

5. Conclusions

1. When the laser powder bed fusion parameters of Ti-Zr-Nb alloy powders vary within the operational borders of TruPrint 1000 systems: the laser power from 40 to 200 W, the laser speed from 200 to 4000 mm/s, the hatching space from 60 to 200 μm, while keeping a constant layer thickness of 30 μm; the higher the laser power density, the lower the titanium and the higher the oxygen and nitrogen contents in the alloy: up to 3.0 at. % loss in titanium and 0.025 and 0.012 (wt.%) gains in oxygen and nitrogen, as compared to the precursor powder.
2. A printed density greater than 99.5% is obtained for specimens printed within the 25 to 110 J/mm^3 laser densities and the 3.2 to 18.7 cm^3/h build rates.
3. The phase state in all the specimens is parent β-phase, with the lattice parameter slightly larger than that of the conventionally processed alloys; this phenomenon is mainly caused by the enrichment of solid solution in Zr and Nb.
4. The lower the laser density energy and the greater the build rate, the finer and the more equiaxed the grain structure of printed specimens, and the less textured the material. In this study, d_{xz} grain size varies from 200 to 50 μm and the grain aspect ratio varies from 4 to 2.
5. Using the experimental data obtained and the dedicated processing simulation algorithm, the density and the grain structure processing maps were built. Discrepancies between the experimentally measured and numerically calculated values do not exceed 2% for the printed density, 15% for the grain size and 11% for the grain aspect ratio.
6. The Ti-Zr-Nb alloy density, grain size and grain aspect ratio processing maps of this study can be used to control the structure of printed specimens and the process productivity, when using a TruPrint 1000 LPBF systems.

Author Contributions: V.B.: project conceptualization, results interpretation, manuscript preparation and editing; V.K.: optical microscopy and data treatment; M.L.: process modeling/calibration and manuscript preparation; K.L.: specimen manufacturing; V.S.: specimen manufacturing, results interpretation and manuscript preparation; S.P.: XRD analysis, results interpretation and editing. All authors have read and agreed to the published version of the manuscript.

Funding: CONMET LLC (Russian Federation). Natural Science and Engineering Research Council of Canada. Russian Science Foundation (project № 20-79-00299).

Acknowledgments: The authors express their gratitude to A. Konopatsky for assistance in chemical analyses and A. Kreitcberg for useful discussions.

Conflicts of Interest: The authors declare that there is no conflict of interest.

References

1. Zafar, M.Q.; Zhao, H. 4D Printing: Future Insight in Additive Manufacturing. *Met. Mater. Int.* **2019**, *26*, 564–585. [CrossRef]
2. Murr, L.E.; Gaytan, S.M.; Medina, F.; Lopez, H.; Martinez, E.; Machado, B.I.; Hernandez, D.H.; Martinez, L.; Lopez, M.I.; Wicker, R.B.; et al. Next-generation biomedical implants using additive manufacturing of complex, cellular and functional mesh arrays. *Philos. Trans. R. Soc. A Math. Phys. Eng. Sci.* **1917**, *368*, 2000–2032. [CrossRef] [PubMed]
3. Bobbert, F.S.L.; Lietaert, K.; Eftekhari, A.A.; Pouran, B.; Ahmadi, S.M. Additively manufactured metallic porous biomaterials based on minimal surfaces: A unique combination of topological, mechanical, and mass transport properties. *Acta Biomater.* **2017**, *53*, 572–584. [CrossRef] [PubMed]
4. *Metal Foams: Fundamentals and Applications*; Dukhan, N. (Ed.) DEStech Publications: Lancaster, PA, USA, 2013; 423p.
5. Sun, S.; Brandt, M.; Easton, M. *Powder Bed Fusion Processes: An Overview*; Elsevier: Amsterdam, The Netherlands, 2017; pp. 55–77.
6. Dadbakhsh, S.; Speirs, M.; Van Humbeeck, J.; Kruth, J.-P. Laser additive manufacturing of bulk and porous shape-memory NiTi alloys: From processes to potential biomedical applications. *MRS Bull.* **2016**, *41*, 765–774. [CrossRef]
7. Ma, J.; Franco, B.; Tapia, G.; Karayagiz, K.; Johnson, L.; Liu, J. Spatial Control of Functional Response in 4D-Printed Active Metallic Structures. *Sci. Rep.* **2017**, *7*, 46707. [CrossRef]
8. Yoneyama, T.; Miyazaki, S. *SMA for Biomedical Applications*; Woodhead: Cambridge, UK, 2008; p. 352.
9. Miyazaki, S.; Kim, H.; Hosoda, H. Development and characterization of Ni-free Ti-base shape memory and superelastic alloys. *Mater. Sci. Eng. A* **2006**, *438*, 18–24. [CrossRef]
10. Kim, H.Y.; Ikehara, Y.; Kim, J.I.; Hosoda, H.; Miyazaki, S. Martensitic transformation, shape memory effect and superelasticity of Ti–Nb binary alloys. *Acta Mater.* **2006**, *54*, 2419–2429. [CrossRef]
11. Li, S.; Cui, T.; Hao, Y.; Yang, R. Fatigue properties of a metastable β-type titanium alloy with reversible phase transformation. *Acta Biomater.* **2008**, *4*, 305–317. [CrossRef]
12. Konopatsky, A.; Dubinskiy, S.; Zhukova, Y.; Sheremetyev, V.; Brailovski, V.; Prokoshkin, S.; Filonov, M. Ternary Ti-Zr-Nb and quaternary Ti-Zr-Nb-Ta shape memory alloys for biomedical applications: Structural features and cyclic mechanical properties. *Mater. Sci. Eng. A* **2017**, *702*, 301–311. [CrossRef]
13. Kudryashova, A.; Sheremetyev, V.; Lukashevich, K.; Cheverikin, V.; Inaekyan, K.; Galkin, S.; Prokoshkin, S.; Brailovski, V. Effect of a combined thermomechanical treatment on the microstructure, texture and superelastic properties of Ti-18Zr-14Nb alloy for orthopedic implants. *J. Alloy. Compd.* **2020**, *843*, 156066. [CrossRef]
14. Prokoshkin, S.; Brailovski, V.; Dubinskiy, S.; Zhukova, Y.; Sheremetyev, V.; Konopatsky, A.; Inaekyan, K. Manufacturing, structure control, and functional testing of Ti–Nb-based SMA for medical application. *Shape Mem. Superelast.* **2016**, *2*, 130–144. [CrossRef]
15. Poulin, J.-R.; Kreitcberg, A.; Terriault, P.; Brailovski, V. Long fatigue crack propagation behavior of laser powder bed-fused inconel 625 with intentionally-seeded porosity. *Int. J. Fatigue* **2019**, *127*, 144–156. [CrossRef]
16. Yang, C.; Zhang, Z.; Li, S.; Liu, Y.-J.; Sercombe, T.; Hou, W.; Zhang, P.; Zhu, Y.; Hao, Y.; Zhang, Z.; et al. Simultaneous improvement in strength and plasticity of Ti-24Nb-4Zr-8Sn manufactured by selective laser melting. *Mater. Des.* **2018**, *157*, 52–59. [CrossRef]
17. Ghosh, S. Predictive modeling of solidification during laser additive manufacturing of nickel superalloys: Recent developments, future directions. *Mater. Res. Express* **2018**, *5*, 012001. [CrossRef]
18. Ahmed, A.; Wahab, M.S.; Raus, A.A.; Kamarudin, K.; Bakhsh, Q.; Ali, D. Effects of Selective Laser Melting Parameters on Relative Density of AlSi10Mg. *Int. J. Eng. Technol.* **2016**, *8*, 2552–2557. [CrossRef]
19. Letenneur, M.; Brailovski, V.; Kreitcberg, A.; Paserin, V.; Bailon-Poujol, I. Laser Powder Bed Fusion of Water-Atomized Iron-Based Powders: Process Optimization. *J. Manuf. Mater. Process.* **2017**, *1*, 23. [CrossRef]
20. Letenneur, M.; Kreitcberg, A.; Brailovski, V. Optimization of Laser Powder Bed Fusion Processing Using a Combination of Melt Pool Modeling and Design of Experiment Approaches: Density Control. *J. Manuf. Mater. Process.* **2019**, *3*, 21. [CrossRef]
21. Kreitcberg, A.; Brailovski, V.; Prokoshkin, S. New biocompatible near-beta Ti-Zr-Nb alloy processed by laser powder bed fusion: Process optimization. *J. Mater. Process. Technol.* **2018**, *252*, 821–829. [CrossRef]

22. Letenneur, M.; Kreitcberg, A.; Brailovski, V. The Average Grain Size and Grain Aspect Ratio in Metal Laser Powder Bed Fusion: Modeling and Experiment. *J. Manuf. Mater. Process.* **2020**, *4*, 25. [CrossRef]
23. Kreitcberg, A.; Brailovski, V.; Sheremetyev, V.; Prokoshkin, S. Effect of Laser Powder Bed Fusion Parameters on the Microstructure and Texture Development in Superelastic Ti–18Zr–14Nb Alloy. *Shape Mem. Superelast.* **2017**, *3*, 361–372. [CrossRef]
24. ASTM Standard. *ASTM E112-10: Standard Test Methods for Determining Average Grain Size*; ASTM International: West Conshohocken, PA, USA, 2010.
25. Schuöcker, D. *Handbook of the Eurolaser Academy*; Springer Science & Business Media: Berilin, Germany, 1998; Volume 2.
26. Kundin, J.; Mushongera, L.; Emmerich, H. Phase-field modeling of microstructure formation during rapid solidification in Inconel 718 superalloy. *Acta Mater.* **2015**, *95*, 343–356. [CrossRef]
27. Nelson, J.B.; Riley, D.P. An experimental investigation of extrapolation methods in the derivation of accurate unit-cell dimensions of crystals. *Proc. Phys. Soc.* **1945**, *57*, 160–177. [CrossRef]
28. Kim, H.; Fu, J.; Tobe, H.; Kim, J.; Miyazaki, S. Crystal structure, transformation strain, and superelastic property of Ti–Nb–Zr and Ti–Nb–Ta alloys. *Shape Mem. Superelast.* **2015**, *1*, 107–116. [CrossRef]
29. Kreitcberg, A.; Sheremetyev, V.; Tsaturyants, M.; Prokoshkin, S.; Brailovski, V. Optimization of Post-processing Annealing Conditions of the Laser Powder Bed-Fused Ti–18Zr–14Nb Shape Memory Alloy: Structure and Functional Properties. *Shape Mem. Superelast.* **2019**, *5*, 172–181. [CrossRef]
30. Gockel, J.; Beuth, J.; Taminger, K. Integrated control of solidification microstructure and melt pool dimensions in electron beam wire feed additive manufacturing of Ti-6Al-4V. *Addit. Manuf.* **2014**, *1–4*, 119–126. [CrossRef]
31. Cherry, J.A.; Davies, H.M.; Mehmood, S.; Lavery, N.; Brown, S.G.R.; Sienz, J. Investigation into the effect of process parameters on microstructural and physical properties of 316L stainless steel parts by selective laser melting. *Int. J. Adv. Manuf. Technol.* **2015**, *76*, 869–879. [CrossRef]
32. Yadroitsau, I. Selective laser melting: Direct manufacturing of 3D-objects by selective laser melting of metal powders. *Lambert Acad. Publ.* **2009**, *55*, 551–560.
33. Popovich, V.; Borisov, E.; Sufiiarov, V.; Masaylo, D.; Alzina, L. Functionally graded Inconel 718 processed by additive manufacturing: Crystallographic texture, anisotropy of microstructure and mechanical properties. *Mater. Des.* **2017**, *114*, 441–449. [CrossRef]
34. Poulin, J.R.; Letenneur, M.; Terriault, P.; Brailovski, V. Influence of Intentionally Induced Porosity and Postprocessing Conditions on the Mechanical Properties of Laser Powder Bed–Fused Inconel 625. *Struct. Integr. Addit. Manuf. Parts* **2018**, *STP1620*, 294–312.
35. Silbernagel, C.; Aremu, A.; Ashcroft, I. Using machine learning to aid in the parameter optimisation process for metal-based additive manufacturing. *Rapid Prototyp. J.* **2019**, *26*, 625–637. [CrossRef]

metals

MDPI

Article

Microstructural and Mechanical Response of NiTi Lattice 3D Structure Produced by Selective Laser Melting

Carlo Alberto Biffi *, Paola Bassani, Jacopo Fiocchi and Ausonio Tuissi

National Research Council—Institute of Condensed Matter Chemistry and Technologies for Energy, Unit of Lecco, CNR ICMATE, Via Previati 1/E, 23900 Lecco, Italy; paola.bassani@cnr.it (P.B.); jacopo.fiocchi@icmate.cnr.it (J.F.); ausonio.tuissi@cnr.it (A.T.)

* Correspondence: carloalberto.biffi@cnr.it; Tel.: +39-0341-2350111

Received: 15 April 2020; Accepted: 9 June 2020; Published: 18 June 2020

Abstract: Nowadays, additive manufacturing (AM) permits to realize complex metallic structural parts, and the use of NiTi alloy, known as Nitinol, allows the integration of specific functions to the AM products. One of the most promising designs for AM is concerning the use of lattice structures that show lightweight, higher than bulk material deformability, improved damping properties, high exchange surface. Moreover, lattice structures can be realized with struts, having dimensions below 1 mm—this is very attractive for the realization of Nitinol components for biomedical devices. In this light, the present work regarded the experimental characterization of lattice structures, produced by selective laser melting (SLM), by using Ni-rich NiTi alloy. Differential scanning calorimetry (DSC), electron backscatter diffraction (EBSD), and compression testing were carried out for analyzing microstructure, martensitic transformation (MT) evolution, and superelasticity response of the SLMed lattice samples. The lattice microstructures were compared with those of the SLMed bulk material for highlighting differences. Localized martensite was detected in the nodes zones, where the rapid solidification tends to accumulate solidification stresses. An increase of martensitic transformation temperatures was also observed in lattice NiTi.

Keywords: shape memory alloys; NiTi; selective laser melting; additive manufacturing; lattice structure; EBSD; superelasticity

1. Introduction

Among the advantages offered by additive manufacturing (AM), the design for freedom is probably the most important because it enables the possibility of integrating extra functionalities, geometrically dependent, into the built parts [1]. In this light, AM of smart materials, including shape memory alloys (SMAs), is considered a challenging research topic. Among SMAs, the most widespread one is the intermetallic Ni–Ti, which exhibits two peculiar properties, namely, shape memory effect and superelasticity [2,3]. These properties have promoted the diffusion of nearly equiatomic NiTi alloy in industrial applications, oriented to actuators and a biomedical field [4,5].

Available literature reports several works based on the study of the microstructure of Nitinol parts, manufactured by the most diffused metal AM technologies, including selective laser melting (SLM), direct energy deposition, and electron beam melting [6]. However, most of the achieved results regard the realization of bulk or massive samples [7,8], while only a few are dedicated to the study of thin or lattice structures in Nitinol [9,10]. It is well known from results regarding other alloys, such as Ti6Al4V and CoCrMo, that production of thin parts or lattice structures is associated with reduced processability windows. Moreover, thin structures can be also more sensitive to the correlation between the thermal gradients, induced during solidification, and the mechanical behavior of the specific alloy.

This effect can be associated with the balance between the generation of internal residual stresses, when the massive geometry of the part does not allow their relaxation, and the thermal distortions, which can occur when the slender geometry permits the stress to relieve.

Therefore, due to the high sensibility of Nitinol to both thermal and mechanical fields [11,12], which both may trigger the martensitic transformation [8,13–16], the investigation of the microstructural properties of Nitinol lattice structures appears as a challenging field. The successful development of such research can lead to the potential use of complex 3D printed parts for industrial applications. In the present work, experimental characterization of Ni-rich NiTi lattice structures, produced by SLM, was carried out. A comparison with SLMed bulk Nitinol samples evidenced that the scanning strategy, adopted for the realization of lattice structures, could promote large variations in microstructure, provoking irregularities in the characteristic phases, typical of this functional alloy. It was found that strong texture could be detected in the nodes of the lattice samples, while the reference bulk samples revealed more uniform microstructure orientation, depending on different thermal histories during the SLM process, which did not cause sensible Ni loss.

2. Experimental

The chosen lattice structures were based on a tetragonal diamond-like geometry (strut diameter and length of 1 mm and 2.5 mm, respectively) and characterized by a relative density of 22% with respect to full dense parts. The samples, having a size of 10 mm × 30 mm × 10 mm (see Figure 1), were produced from Ni50.8Ti49.2 (at.%) powder, having a size approximately in the range 5–50 µm, with an SLM system (mod. AM400 from Renishaw, New Mills, Kingswood, Wotton-under-Edge GL12 8JR, UK). The main process parameters, which were investigated elsewhere [17], are listed in Table 1. As a reference, bulk samples (6 mm in height and 3 mm in diameter) were printed using equivalent parameters. Lattice and bulk samples were realized with contour and meander scanning strategy, respectively.

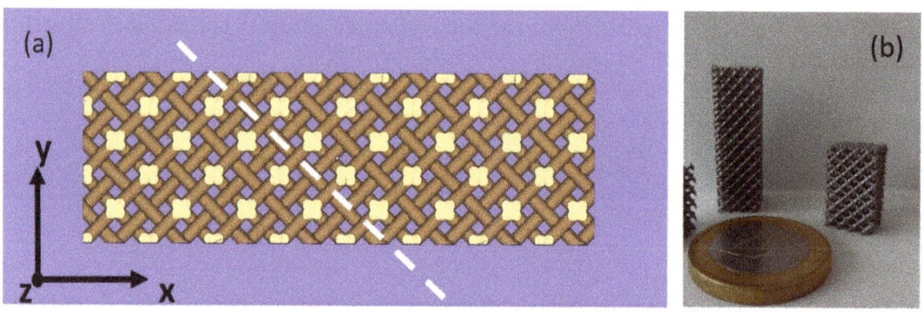

Figure 1. Schematic of the lattice structure, built onto the building platform (**a**), highlighting the building direction (z) containing section where electron backscatter diffraction (EBSD) analysis was performed, and NiTi lattice samples (**b**).

Table 1. Process parameters used for printing Nitinol lattice structure and bulk samples.

Parameters	Values
Power	75 W
Exposure time	75 µs
Atmosphere	Argon
Layer thickness	30 µm
Hatch distance	50 µm
Point distance	50 µm
Laser spot size	65 µm

X-ray computed tomography (CT) was performed on a prismatic portion of a sample, using an XTH225–ST system (Nikon, Leuven, Belgium), having an x-ray gun of 225 kV and a 16-bit flat-panel Varex 4343CT as a detector, to highlight the defects within the entire volume of the sample. The following settings were used for the measurements: (i) voltage of 154 kV; (ii) 7 mm as the resolution of the detector; (iii) 105 min as scanning time.

The samples were characterized in the as-built condition. Differential scanning calorimetry (DSC) was performed on small specimens, about 30 mg in weight, cut with a diamond saw from the built parts: a complete thermal cycle in the (−100 °C; 150 °C) temperature range with heating/cooling rate of 10 °C/min was carried out using a DSC (mod. SSC 5200 by Seiko Instruments, Chiba, Japan). The microstructure was investigated through scanning electron microscopy, coupled with electron backscatter diffraction (EBSD) and energy-dispersive x-ray spectroscopy (EDXS) on the XY view for both the samples (FEG-SEM SU70 from Hitachi, Chiyoda, Japan, equipped with EBSD and EDXS systems from Thermo Scientific, Waltham, MA, USA); moreover, the lattice structure was also analyzed in a section parallel to the building (Z) direction and lying at 45° with respect to the XY plane, as reported in Figure 1 in dashed line. Sections were included in graphite-loaded hot mounting resin, and metallographic surfaces with a final colloidal silica finishing were prepared. Mechanical uniaxial compression tests were performed by means of an MTS 2/M machine (MTS Systems Corporation, Eden Prairie, MN, USA), equipped with an extensometer, at a strain rate of 0.01 min^{-1}. Eight complete loading and unloading cycles, up to 10% in strain, were carried out at room temperature (RT). Prior to testing, all the samples were heated up to 70 °C and then cooled to RT to allow the characterization of the samples in the same condition and to evaluate recoverable strain after unloading.

3. Results and Discussion

A CT scan performed on the lattice structure for analyzing the internal defects is reported in Figure 2. In detail, a representative cross-section of the lattice structure, observed along the XY plane, is shown in Figure 2a. It can be seen that the porosity had no preferential distribution within the nodes. This fact should depend on the scanning strategy adopted during the SLM process: the contour strategy implied an ellipsoidal path of the laser beam from the center to the border of the lattice element; therefore, the best degree of homogeneity on the overlapping of adjacent liquid pools could be reached, as also reported in other works [18]. The analysis of the defects revealed that the defect size ranged between 30 μm and 250 μm, except few larger defects (see Figure 2b), and the sphericity of the main fraction of the defects was in the 0.45–0.7 range. As discussed in previous works [17,19], this behavior suggested that the energy irradiated by the laser beam to the powder bed was sufficient for producing high-density samples.

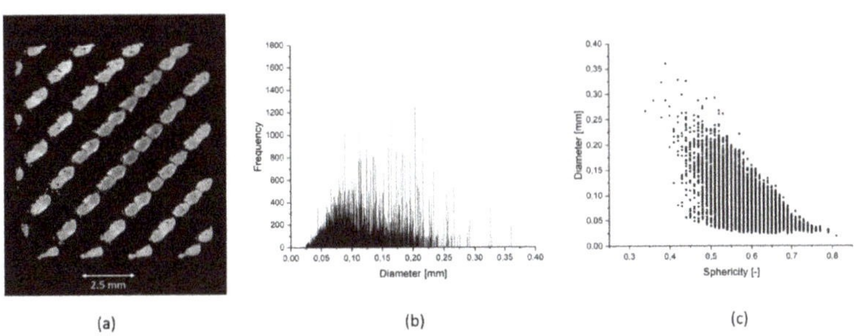

Figure 2. Computed tomography (CT) analysis of the NiTi SLMed structure: cross-section in the XY view (**a**), frequency- diameter (**b**), and diameter-sphericity (**c**) trend.

DSC scans of the lattice and bulk samples are shown in Figure 3, while the characteristic temperatures of the MT (austenite start, peak and finish temperatures, As, Ap and Af, and martensite start, peak and finish temperatures, Ms, Mp and Mf) and the corresponding heats exchanged ($H_{A \to M}$ and $H_{M \to A}$) are listed in Table 2.

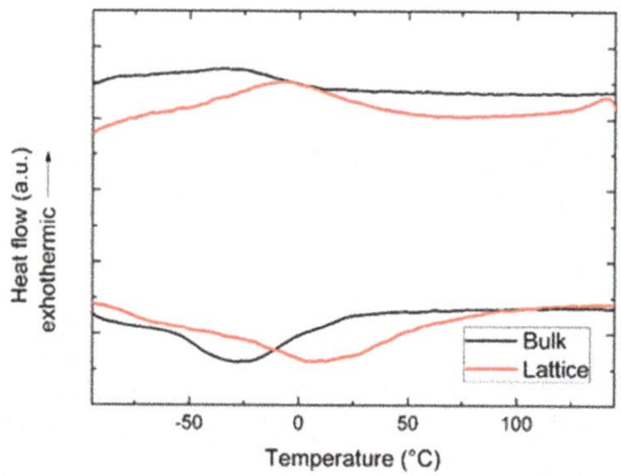

Figure 3. DSC (differential scanning calorimetry) scans of the lattice and bulk samples.

Table 2. Characteristic temperatures of the martensitic transformation (MT) measured for the lattice and bulk samples.

Sample:	As (°C)	Ap (°C)	Af (°C)	$H_{M \to A}$ (J/g)	Ms (°C)	Mp (°C)	Mf (°C)	$H_{A \to M}$ (J/g)
Lattice	−47	9	76	10.5	44	−5	−52	7.5
Bulk	−61	−22	30	5.4	11	−28	−86	4.0

Due to the same fluence being used during the SLM process (127 J/mm^3), the increase of temperatures of the martensitic transformation in the lattice sample should reasonably due to the presence of residual stresses: this was in good agreement with the Clausius–Clapeyron equation for shape memory alloys. In fact, the lattice structure was realized through a contour scanning strategy, which was dedicated to thin parts, but also caused higher heat accumulation, due to the laser scanning on an almost circular path. On the contrary, the bulk sample was produced using the meander scanning strategy, which promoted subsequent heating/cooling cycles when the laser beam moved along parallel lines. More intense heat accumulation could lead to slower cooling rates; therefore, grain growth was favorite. Additionally, it could be also mentioned that higher heat accumulation could favor Ni loss, as well as the generation of residual stresses, which are typical of the SLM process, and this resulted in an increase in the temperatures of the MT.

EBSD analyses performed on the XY section of the two specimens are reported in Figure 4. Indexing was performed according to the body-centered cubic austenite. Orientation image micrographs (OIM), calculated for the normal direction of the section (i.e., the building direction (BD) for XY section like the proposed ones), inverse pole figures (IPF), calculated for the normal direction, and (100) pole figures (PF) were reported, respectively. Two main differences between lattice and bulk specimen were well apparent.

The lattice showed (see Figure 4a) the presence of not-indexed (black areas), that could be ascribed to localized martensite, and a not uniform texture; the latter visualized as a more reddish center and randomly colored border regions in OIM of Figure 4b. On the contrary, the bulk specimen (see Figure 4d) showed a more homogeneous microstructure, with a faint (100) texture aligned with BD.

The non-homogeneous texture of the lattice specimen is highlighted in Figure 5. IPFs calculated on the two areas demonstrated that a sharp (100) texture parallel to the BD was present in the center of the section, while a more random texture was observed near the contour. Analyses performed on BD sections and at higher magnification allowed to partly clarify the origin of both not-indexed regions and texture differences. Columnar grains with (100) texture were mostly developed in the middle of the nodes that had the shape of ligaments built parallel to BD. Contour grains in the struts tended to grow in a more bent direction due to the different spatial thermal gradient experienced by the material in these regions. Equiaxed grains were also present, whose formation could be ascribed to the different intensity of the thermal gradient with respect to the internal regions. A different heat extraction close to the borders was likely due to the surrounding powder—not melted by the laser beam—which had a thermal conductivity lower than the consolidated material.

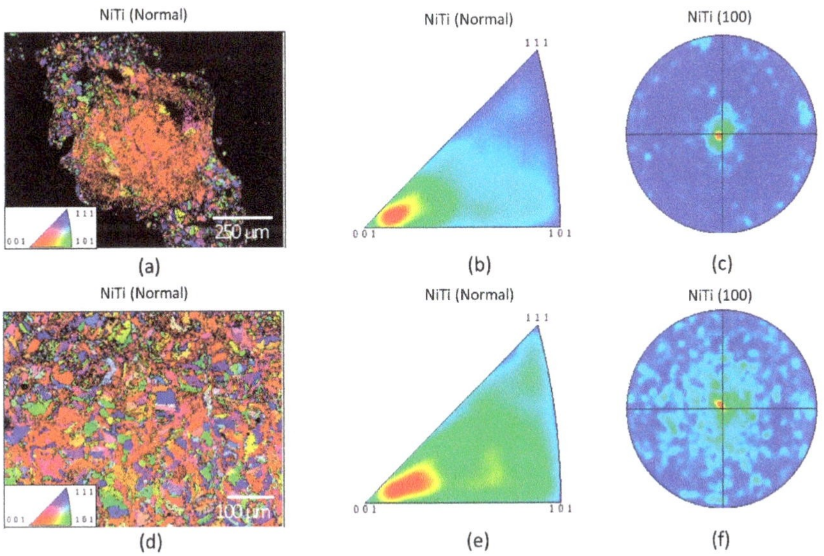

Figure 4. EBSD results on XY section of the lattice (**a**–**c**) and bulk (**d**–**f**) samples, analyzed with orientation image micrographs (OIM) related to building direction (BD), inverse pole figures (IPF), and pole figures (PF) of the analyzed areas, respectively.

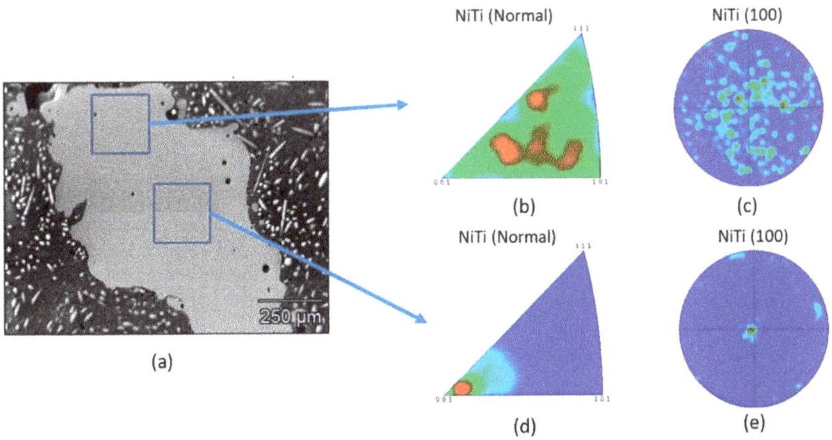

Figure 5. SEM-BSE micrograph acquired on the XY section (**a**). IPF and PF of two regions were reported: (**b**,**c**) region close to the border and (**d**,**e**) central part of the lattice structure.

Not indexed regions, i.e., martensitic areas (see Figure 6c), seemed to well correspond to stress concentration regions, namely, the connection volume between the struts and nodes, as well as the upper surface of nodes. Consequently, they possibly correspond to the presence of bands of stress-induced martensite (SIM) [2,3]. This hypothesis was supported by the fact that the solidification stresses were accumulated in the joint between two or more bridges (massive parts). On the contrary, the extreme parts of the lattice structure, standing alone and free to deform, were able to relax the stresses induced during the rapid solidification, and they reasonably did not exhibit any SIM, as shown in Figure 6. In fact, EBSD performed on the lattice part indicated that only austenite could be detected, and no martensite was present in the formed columnar struts.

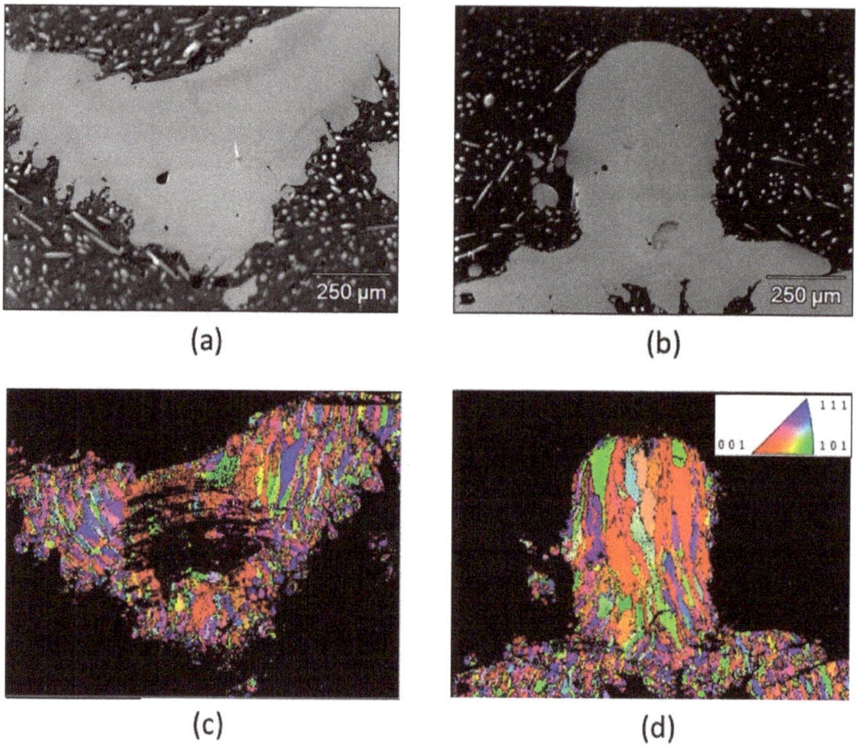

Figure 6. EBSD analyses on the BD section of lattice, involving a node and concurrent struts (**a**,**c**) and an extreme part of the lattice structure (**b**,**d**). a, b Secondary electron reference micrographs, c, d—OIM related to BD.

Moreover, higher magnification analyses in the lattice structure revealed the ideal continuity of austenite grains above and below martensitic regions, as represented in Figure 7. EDX analyses of lattice, performed on areas of about 200 µm × 200 µm, didn't reveal any statistically significant compositional differences. On the contrary, a small variation in Ni content was observed at the micrometric scale (see Figure 7d–f).

It is worth noting that, in this case, the EDXS analyses were indeed performed close to their detectability limit. All these considerations led to the conclusion that wide areas of stress-induced martensite were present in the lattice structure. In terms of comparison, the compositional analysis performed on the bulk sample revealed that Ni content was 50.9 ± 0.15, measured in atomic percentage, and the average Ni content in the lattice samples was 50.95%; thus, no significant compositional variations were detected between lattice and bulk specimens due to the use of equivalent process parameters [20].

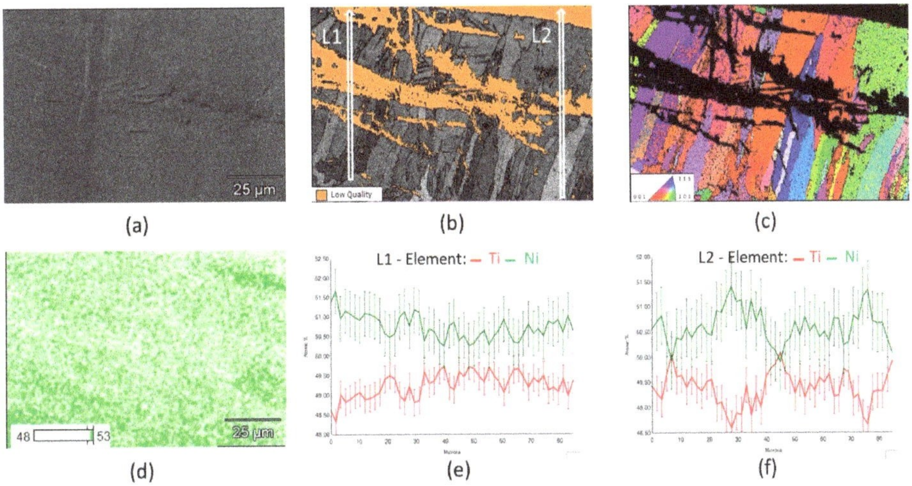

Figure 7. Higher magnification EBSD/EDXS analysis on BD section: (**a**) reference secondary electron micrograph, (**b**) pattern quality map, (**c**) OIM related to BD, (**d**) EDXS Ni distribution map, (**e**,**f**) extracted line analyses of line 1 and 2 in Figure 6b.

The mechanical behavior of the Nitinol lattice structure under subsequent loading/unloading cycles under compression is shown in Figure 8a. In terms of comparison, the mechanical behavior of the bulk sample tested in compression, thoroughly described in [17], is shown in Figure 9. During deformation, the reached stresses were obviously much lower than the ones which could be expected in a bulk sample since they were computed considering the sample as if it was "fully dense", according to the widespread agreement [21]. Recovered strains (Figure 8c) increased almost linearly with imposed deformation. In particular, an interesting value of 4.5% was reached for a 10% imposed deformation. At large deformations (higher than 7%), the recovery abilities inherent to the geometry's low stiffness were likely to start playing an overwhelming role. At the same time, the beginning of the lattice densification gave rise to a plateau-like behavior. It should be noted that the bulk samples appeared to allow a higher degree of recovery, at least up to 7% deformation. This behavior, which might seem to be counterintuitive at first, was explained by considering that, as highlighted by EBSD analysis, trabecular structures contained a higher amount of martensite in the as-built state. Moreover, it was likely that the stress concentration acting at each node of the structure was able to induce martensite at early stages and readily plastically deform it, thus preventing the reverse transformation into austenite. In order to have a complete perspective on the mechanical properties of the produced NiTi lattice structures, a sample was deformed in compression up to 4% and 6% and then heated to approximately 80 °C. These temperatures, as shown in Figure 2, laid above the A_f temperature and were, therefore, expected to induce the transformation of the SIM into austenite. In fact, both residual deformations upon unloading (1.72% and 2.75%, respectively) were completely recovered after heating at 80 °C (see Figure 8c).

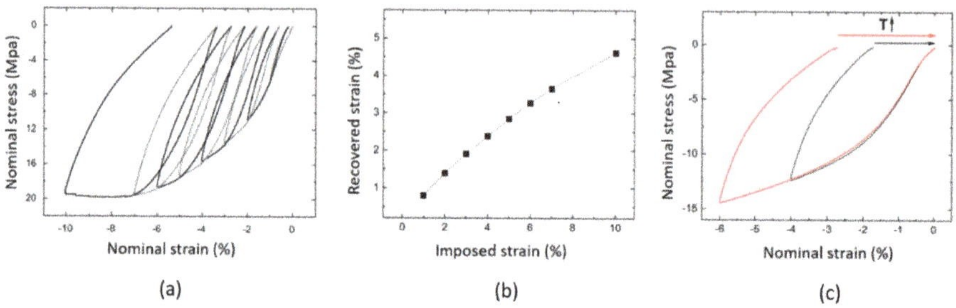

Figure 8. Mechanical behavior under compression of the lattice sample (**a**); the trend of recovered vs. imposed deformation (**b**); strain recovery (**c**).

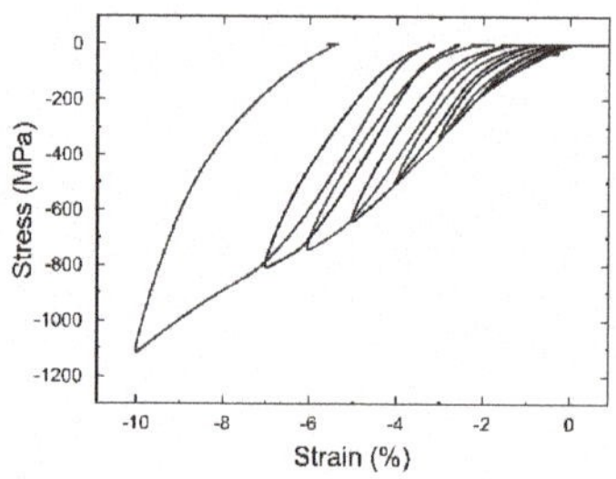

Figure 9. Mechanical behavior under compression of the bulk sample [18].

4. Conclusions

In this work, the microstructure and the mechanical behavior of Nitinol lattice structures, produced by selective laser melting, were studied. It can be highlighted that:

(1) The martensitic transformation of the lattice structure occurred at temperatures slightly higher than the bulk material due to the formed residual stresses;

(2) The lattice structures exhibited nodes with some areas of stress-induced martensite and fully austenitic struts, while the bulk sample was fully austenitic;

(3) Grains were oriented along preferential directions in the lattice structures, while they were more casually oriented in the bulk sample;

(4) Compression behavior of the lattice structure indicated that the recovered strain increased directly proportionally with the imposed strain; moreover, at room temperature, the lattice structure exhibited a partial superelastic response, while residual deformation was recovered after heating up to 80 °C.

Author Contributions: Conceptualization, C.A.B. and A.T.; methodology, C.A.B.; validation, all authors.; formal analysis, C.A.B., J.F. and P.B.; investigation, C.A.B., J.F. and P.B.; resources, A.T.; writing—original draft preparation, C.A.B., J.F. and P.B.; writing—review and editing, all authors; visualization, C.A.B.; supervision, A.T.; project administration, A.T.; funding acquisition, A.T. All authors have read and agreed to the published version of the manuscript.

Funding: This research was funded through Accordo Quadro between CNR and Regione Lombardia, grant number 3866.

Acknowledgments: The authors would like to acknowledge Saes Getters SpA as a supplier of Nitinol powder, Nikon for the support in the Computer Tomography analysis, and Accordo Quadro CNR/Regione Lombardia n. 3866 at 17/07/2015 FHfFC for financial support. The authors would like to thank Nicola Bennato from CNR ICMATE for his support in the experimental activity.

Conflicts of Interest: The authors declare no conflict of interest.

References

1. Tang, Y.; Zhao, Y.F. A survey of the design methods for additive manufacturing to improve functional performance. *Rapid Prototyp. J.* **2016**, *22*, 569–590. [CrossRef]
2. Otsuka, K.; Wayman, C.M. *Shape Memory Materials*; Cambridge Univ. Press: Cambridge, UK, 1998.
3. Otsuka, K.; Ren, X. Physical Metallurgy of Ti-Ni based shape memory alloys. *Prog. Mater. Sci.* **2005**, *50*, 511–678. [CrossRef]
4. Van Humbeeck, J. Non medical applications of shape memory alloys. *Mater. Sci. Eng. A* **1999**, *273*, 134–148. [CrossRef]
5. Duerig, T.; Pelton, A.; Stöckel, D. An overview of nitinol medical applications. *Mater. Sci. Eng. A* **1999**, *273*, 149–160. [CrossRef]
6. Elahinia, M.; Shayesteh Moghaddam, N.; Taheri Andani, M.; Amerinatanzi, A.; Bimber, B.A.; Hamilton, R.F. Fabrication of NiTi through additive manufacturing: A review. *Prog. Mater. Sci.* **2016**, *83*, 630–663. [CrossRef]
7. Saedi, S.; Moghaddam, N.S.; Amerinatanzi, A.; Elahinia, M.; Karaca, H.E. On the effects of selective laser melting process parameters on microstructure and thermomechanical response of Ni-rich NiTi. *Acta Mater.* **2018**, *14*, 552–560. [CrossRef]
8. Speirs, M.; Wang, X.; van Baelen, S.; Ahadi, A.; Dadbakhsh, S.; Kruth, J.P.; Van Humbeck, J. On the Transformation Behavior of NiTi Shape-Memory Alloy Produced by SLM. *Shape Mem. Superelasticity* **2016**, *2*, 310. [CrossRef]
9. Dadbakhsh, S.; Speirs, M.; Van Humbeck, J.; Kruth, J.P. Laser additive manufacturing of bulk and porous shape memory NiTi alloys: From processes to potential biomedical applications. *MRS Bull.* **2016**, *41*, 765–774. [CrossRef]
10. Andani, M.T.; Saedi, S.; Turabi, A.S.; Karamooz, M.R.; Haberland, C.; Karaca, H.E.; Elahinia, M. Mechanical and Shape Memory properties of porous Ni50.1Ti49.9 alloys manufactured by selective laser melting. *J. Mech. Behav. Biomed. Mater.* **2017**, *68*, 224–231. [CrossRef] [PubMed]
11. Fiocchi, J.; Biffi, C.A.; Scaccabarozzi, D.; Saggin, B.; Tuissi, A. Enhancement of the Damping Behavior of Ti6Al4V Alloy through the Use of Trabecular Structure Produced by Selective Laser Melting. *Adv. Eng. Mater.* **2019**. [CrossRef]
12. Demir, A.G.; Previtali, B. Additive manufacturing of cardiovascular CoCr stents by selective laser melting. *Mater. Des.* **2017**, *119*, 338–350. [CrossRef]
13. Funakubo, H. *Shape Memory Alloys*; Gordon and Breach Science Publishers: Amsterdam, The Netherlands, 1987.
14. Nishida, M.; Wayman, C.; Honma, T. Precipitation processes in near-equiatomic TiNi shape memory alloys. *Metall. Trans. A* **1986**, *17*, 1505–1515. [CrossRef]
15. Mahmoudi, M.; Tapia, G.; Franco, B.; Ma, J.; Arroyave, R.; Karaman, I.; Elwany, A. On the printability and transformation behavior of nickel-titanium shape memory alloys fabricated using laser powder-bed fusion additive manufacturing. *J. Manuf. Process.* **2018**, *35*, 672–680. [CrossRef]
16. Bormann, T.; Muller, B.; Schinhammer, M.; Kessler, A.; Thalmann, P.; de Wild, M. Microstructure of selective laser melted nickel-titanium. *Mater. Charact.* **2014**, *94*, 189–202. [CrossRef]
17. Biffi, C.A.; Fiocchi, J.; Bassani, P.; Tuissi, A. Selective Laser Melting of NiTi Shape Memory Alloy: Processability, microstructure and superelasticity. *Shape Mem. Superelasticity* **2020**. Under consideration.
18. Biffi, C.A.; Fiocchi, J.; Ferrario, E.; Fornaci, A.; Riccio, M.; Romeo, M.; Tuissi, A. Effects of the scanning strategy on the microstructure and mechanical properties of a TiAl6V4 alloy produced by electron beam additive manufacturing. *Int. J. Adv. Manuf. Technol.* **2020**, 1–12. [CrossRef]
19. Fiocchi, J.; Biffi, C.A.; Tuissi, A. Selective laser melting of high-strength primary AlSi9Cu3 alloy: Processability, microstructure, and mechanical properties. *Mater. Des.* **2020**. [CrossRef]
20. Shiva, S.; Palani, I.A.; Mishra, S.K.; Paul, C.P.; Kukreja, L.M. Investigation on the influence of composition in the development of Ni-Ti shape memory alloy using laser based additive manufacturing. *Opt. Laser Technol.* **2015**, *69*, 44–51. [CrossRef]
21. Ashby, M.F.; Evans, A.G.; Fleck, N.A.; Gibson, L.J.; Hutchinson, J.W. *Metal Foams: A Design Guide, Butterworth-Heinemann*; United States of America: Boston, MA, USA, 2000.

metals

MDPI

Article

Softening of Shear Elastic Coefficients in Shape Memory Alloys Near the Martensitic Transition: A Study by Laser-Based Resonant Ultrasound Spectroscopy

Petr Sedlák [1], Michaela Janovská [1], Lucie Bodnárová [1], Oleg Heczko [2,3] and Hanuš Seiner [1,*]

1 Institute of Thermomechanics, Czech Academy of Sciences, Dolejškova 5, 182 00 Prague, Czech Republic; psedlak@it.cas.cz (P.S.); janovska@it.cas.cz (M.J.); bodnarov@it.cas.cz (L.B.)
2 Institute of Physics, Czech Academy of Sciences, Na Slovance 1999/2, 182 21 Prague, Czech Republic; heczko@fzu.cz
3 Faculty of Mathematics and Physics, Charles University, Ke Karlovu 5, 121 16 Prague, Czech Republic
* Correspondence: hseiner@it.cas.cz; Tel: +42-026-605-3672

Received: 15 September 2020; Accepted: 07 October 2020; Published: 16 October 2020

Abstract: We discuss the suitability of laser-based resonant ultrasound spectroscopy (RUS) for the characterization of soft shearing modes in single crystals of shape memory alloys that are close to the transition temperatures. We show, using a numerical simulation, that the RUS method enables the accurate determination of the c' shear elastic coefficient, even for very strong anisotropy, and without being sensitive to misorientations of the used single crystal. Subsequently, we apply the RUS method to single crystals of three typical examples of shape memory alloys (Cu-Al-Ni, Ni-Mn-Ga, and NiTi), and discuss the advantages of using the laser-based contactless RUS arrangement for temperature-resolved measurements of elastic constants.

Keywords: shape memory alloys; martensitic transitions; phonon softening; resonant ultrasound spectroscopy; laser-ultrasound; elastic constants

1. Introduction

Athermal martensitic transitions are known to be mediated by soft acoustic phonons [1–3]. In the shape memory alloys (SMAs), the relation between the structural change and acoustic phonons can be most apparently observed for the forward transition, i.e., the transition from the high temperature phase (austenite) to either the low temperature phase (martensite) or some intermediate phases (R-phase in NiTi [4], premartensite in stoichiometric Ni_2MnGa [5], and similar). Prior to this transition itself, the austenite crystal lattice anomalously softens with respect to specific shearing modes when approaching the transition temperature [6–8], which indicates acoustic phonon condensation. Most typically, for materials for which the spontaneous lattice strain due to the transition is equivalent to tetragonal distortion along the the Bain path, the softening is observed for the $\langle 1\,1\,0\rangle\{1\,\bar{1}\,0\}$ shears in the cubic lattice (i.e., the TA_2 phonon).

In terms of elastic constants of cubic austenite, this shear softening can be observed as a decrease of the elastic constant c', which corresponds to the slope of the TA_2 phonon branch in the low-frequency limit. Cubic materials have three independent elastic constants: c_{11}, c_{12}, and c_{44} (in the Voigt notation), and the shear elastic constant c' can be expressed as

$$c' = \frac{c_{11} - c_{12}}{2}. \tag{1}$$

As the instability of the lattice increases when approaching the martensitic transition temperature, this constant becomes much softer than the constant c_{44} representing the stiffness against basal ($\langle 1\,0\,0\rangle\{0\,1\,0\}$) shears. Hence, the Zener anisotropy ratio (e.g., [9])

$$A = \frac{c_{44}}{c'} = \frac{2c_{44}}{c_{11} - c_{12}} \tag{2}$$

can be used as an appropriate scalar parameter for quantifying the instability. While for single crystals of pure metals this ratio is typically between 1 and 3 (3.21 for copper is already considered to be a strong anisotropy [10]), for SMAs the values of A reach often above 10, or even higher ($A \approx 25$ was reported for Co-Ni-Al single crystals in [11], and up to $A \approx 35$ can be reached for Ni-Fe-Ga-based ferromagnetic SMAs [12]). This extremely strong anisotropy puts some specific requirements on the experimental methods that are used for the determination of the elastic coefficients. Traditionally, two concurrent ultrasonic methods are used for this purpose: the time-of-flight method (TOF, using either through-transmission or pulse-echo arrangements, e.g., [13]) and the resonant ultrasound spectroscopy (RUS [14,15]). In the TOF-based measurements, the elastic constant is determined from a velocity of bulk acoustic waves propagating in a given direction of the material; in particular, the elastic constant c' can be calculated from the velocity v' of the $[1\,\bar{1}\,0]$−polarized (i.e., the slowest) shear wave propagating in the $[1\,1\,0]$ direction, as

$$c' = \rho(v')^2, \tag{3}$$

where ρ is the mass density. The TOF-based measurements are typically performed while using an ultrasonic transducer directly contacted to one face of the sample. The transducer emits planar waves propagating in direction perpendicular to the face, which are then recorded at either the opposite face (the through-transmission arrangement) while using another transducer, or by the same transducer after reflecting from the opposite face (the pulse-echo arrangement). However, the TOF-based measurements can be also performed while using contact-less methods, such as electromagnetic-acoustic transducers (EMAT, [16]), as utilized e.g., for the TOF-based characterization of the elastic softening of the Ni-Mn-Ga SMA by Vasil'ev et al. [17].

In the RUS method, the elastic constants are determined from the resonant frequencies of free elastic vibrations of a small (typically parallelepiped-shaped) sample. Because the relation between the resonant frequency and the elastic constants cannot be explicitly expressed by any formula similar to (3), a numerical simulation of the vibrations is needed, and the elastic constants are, using this simulation, obtained as a result of an inverse problem solution, i.e., by minimizing the misfit between the measured resonant frequencies and those calculated for iteratively improved guesses of elastic constants. The number of resonant frequencies utilized for the inverse procedure may vary from less than ten to nearly one hundred, depending on the vibration damping in the given material; hence, in principle, a more complex information on the elasticity of the material can be obtained by a single measurement than from the TOF-based approaches.

Both of these methods have their advantages and disadvantages and they can be shown to complement each other [18,19]. Nevertheless, for determination of c' for strongly anisotropic cubic materials, which is the most important task for SMA single crystals, the RUS approach appears to be the more suitable one. The main reason is that the TOF-based approaches are very sensitive to the exact knowledge of the orientation of the crystal that was used for the measurements. In Figure 1, this sensitivity is illustrated using a simple numerical simulation: we assumed a rectangular single crystal of dimensions $3 \times 3.5 \times 4$ mm^3 suitable for TOF measurements of v', i.e., with one edge oriented along the $[1\,1\,0]$ direction. For such a crystal, the error in c' calculated via eqrefcprimev when the crystal was slightly misoriented by some angle ψ was calculated. The calculation was done using Christoffel's equation for wave propagation in a general direction of an anisotropic material [9], and assuming that c_{11} and c_{44} were known and fixed (the values $c_{11} = 140$ GPa and $c_{44} = 60$ GPa were taken for the simulation). Figure 1a presents the result. For as low misorentation as $\psi \approx 2°$, an up to 10% error can arise for strong anisotropy, while for $\psi \approx 5°$ the error goes already above 50%.

Notice that the misorientation of the direction along which the measured ultrasonic waves propagate may originate not only from an imprecise crystal cutting or the experimental error of the Laue/EBSD characterization techniques. The $[1\,1\,0]$ direction in cubic materials with $A > 1$ is energetically shaded for $[1\,\bar{1}\,0]$-polarized shear waves [20], i.e., the energy of waves propagating in the material tend to be driven away from this direction, unless a perfectly planar, spatially infinite source of waves is considered. Hence, for ultrasonic probes with a finite aperture (i.e., an aperture comparable to the propagation distance), the detected waves with the highest amplitude can correspond to slightly inclined propagation directions.

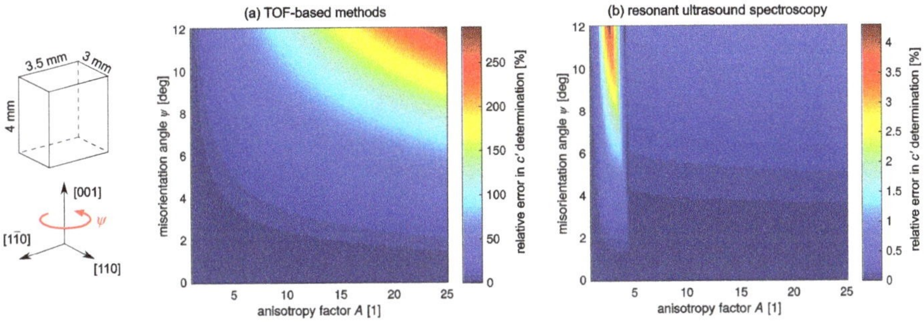

Figure 1. Simulated errors in c' resulting from a small misorentation of the sample: (**a**) methods based on time-of-flight measurements; (**b**) resonant ultrasound spectroscopy. The misorientation angle ψ is defined in the sketch on the very left.

Figure 1b shows a similar simulation for RUS, assuming that the c' coefficient was determined from the lowest resonant mode of the sample (it was checked that involving up to 10 resonant modes in the simulation leads to very similar results). The inverse calculation of the elastic constant from the resonant frequency was done while using the Ritz–Rayleigh method-based approach described in [21]. Here, the highest error in c' is obtained for relatively small anisotropy factors ($A \approx 3$) and high misorientation angles, but it never exceeds 5%. However, for very strong anisotropy, the error stays below 1%, even for significant misorientations. The reason is that the lowest resonant modes used in RUS are always those that are related to the softest shears of the lattice, i.e., for $A \to \infty$ these modes are always purely $\langle 1\,1\,0\rangle\{1\,\bar{1}\,0\}$ shearing modes, regardless of the exact orientation of the sample. For moderate anisotropy factors, the lowest modes are still also somehow affected by the c_{44} shear coefficient, and the partitioning between the effects of c' and c_{44} coefficients changes with the misorientation. Hence, the RUS measurement for an incorrectly oriented single crystal with $A \approx 3$ may result in obtaining a mixture of c' and c_{44} instead of pure c'.

Besides this genuine insensitivity to the sample misorientation, the RUS method possesses several additional features that predetermine it for the characterization of SMAs. When using lasers for both generation and detection of the vibrations [21–23], the sample can be kept in exactly stress-free conditions (no clamping force from the ultrasonic transducers) and, thus, the observed thermally induced transitions are not biased by a presence of any mechanical loads. Furthermore, the laser-based arrangement allows for the sample to be put into a temperature chamber, which increases the accuracy of the temperature control and allows the measurements to be performed in a broader range of temperatures. The contact-less arrangement also enables an accurate assessment of the internal friction parameters from the RUS measurements [24,25], where the internal friction Q^{-1} can be directly determined from the widths of the resonant peaks (see Section 2). This is the parameter only hardly accessible by the TOF-based methods. Finally, the RUS measurements allow, in principle, also the harder shear elastic constant (c_{44}) to be determined from the same measurement, assuming that a sufficient number of resonant frequencies is involved in the analysis and/or that the anisotropy factor is not extremely high.

In this paper, we show several examples of laser-based RUS measurements on SMA single crystals near the martensitic transformation, exploiting the above-listed advantages. For a Cu-Al-Ni single crystal, we utilize the ability of RUS to determine accurately the c' coefficient in a narrow temperature interval close to the transition temperature and discuss the correlation between the elastic softening and the internal friction; for Ni-Mn-Ga single crystals with two different chemical compositions, we show the advantage of high-temperature RUS measurements for materials with extremely strong anisotropy, and finally for a Ni-Ti single crystal we demonstrate the ability of RUS to simultaneously monitor the evolutions of c' and c_{44}.

2. Materials and Methods

Four single crystals of different SMAs were used for the study. All of the samples were rectangular bars with one face mirror-polished to enable laser based detection of the vibrations; for all samples, the crystallographic orientations of the individual faces were determined by Laue method and the mass density was determined by Archimedes method. The transition temperatures were either determined by differential scanning calorimetry (samples Nos. 1, 2, and 4), or estimated directly from the changes of the elastic constants (sample No.3). Hereafter, we will use the common notation of the transition temperatures, with A_S standing for the austenite start temperature, A_F for the austenite finish temperature, and similarly for martensite (M_S and M_F) and the R-phase in NiTi (R_S, R_F); these temperatures for all samples are summarized in Table 1, where the chemical compositions of the individual samples are also listed. The samples were:

Table 1. Chemical compositions of the studied samples and the corresponding transition temperatures determined by differential scanning calorimetry (if not stated otherwise).

Sample No.	Alloy	Composition (at.%)	A_S [K]	A_F [K]	M_S [K]	M_F [K]	R_S [K]	R_F [K]
1	Cu-Al-Ni	$Cu_{69.4}Al_{27.2}Ni_{3.4}$	395	314	299	284	-	-
2	Ni-Mn-Ga	$Ni_{50.0}Mn_{28.9}Ga_{21.1}$	328	329	328	326	-	-
3	Ni-Mn-Ga	$Ni_{50.5}Mn_{30.4}Ga_{19.1}$	400 [a]	420 [a]	-	-	-	-
4	Ni-Ti	multiphase [b]	240	286	232	226	~270	245

[a] estimated from RUS measurements; [b] consisting of a nearly equiatomic matrix with Ti_2Ni and TiC particles; total averaged composition $Ni_{48.8}Ti_{51.2}$ (at %).

1. a Cu-Al-Ni single crystal (sample dimensions in austenite 3.05 × 2.03 × 1.29 mm^3, sample orientation in austenite approximately $[1\,1\,0] \perp [1\,\bar{1}\,0] \perp [0\,0\,1]$, mass density 6.78 g·cm^{-3}). This sample was characterized with cooling from 330 K towards the martensite start temperature (M_S, 299 K). To discern sufficient details of the behavior near the transition temperature, the measurement was done with a 0.5 K step (temperature control ±0.01 K) and the internal friction parameter Q^{-1} was evaluated at each temperature. Six lowest resonant modes were used for the determination of c' and the analysis of the internal friction;
2. a Ni-Mn-Ga single crystal (sample dimensions in austenite 3.41 × 2.80 × 0.92 mm^3 at 335 K, sample orientation in austenite approximately $[1\,0\,0] \perp [0\,1\,0] \perp [0\,0\,1]$, mass density 8.12 g·cm^{-3}). This sample was characterized with cooling from above the Curie temperature (T_C, 372 K) towards $M_S = 328$ K, where the sample transformed into 10 M modulated martensite [26] (a thermally induced mixture of variants). Afterwards, an external magnetic field was used for martensite reorientation [27,28], and the sample was set in a single (tetragonal) variant state with the $c-$axis that was oriented along the shortest edge. Note that the 10 M martensite is slightly monoclinic, so this state was still a mixture of monoclinic variants sharing a common $c-$axis, but such a microstructure can be understood as a single variant in the tetragonal approximation [29]. The sample in this condition was then characterized with heating from 289 K until the sample

transformed back to austenite (austenite start temperature, $A_S \approx M_S$). This procedure was repeated second time, with the $c-$axis set along the longest edge of the sample. For the sample in austenite, ten lowest resonant modes were used for c' determination. However, for temperatures that were below M_S, the strong internal friction in 10 M martensite did not allow for reliable detection and identification of more than two or three resonant modes. This was insufficient for the reliable determination of any of the elastic coefficients of the (effectively tetragonal) mixture of monoclinic martensites. Hence, instead, we analyzed the RUS data assuming again cubic symmetry, which resulted in some effective c' for the martensite. Although this effective elastic constant has no direct relation to any shearing mode of the 10 M modulated lattice, its temperature evolution provides a qualitative information on the shear instability in martensite (see e.g., [30,31] for similar approaches);

3. another Ni-Mn-Ga single crystal (dimensions in martensite $4.92 \times 4.65 \times 3.40$ mm^3, sample orientation approximately $[0\,0\,1]_{\mathrm{bct}} \perp [0\,1\,0]_{\mathrm{bct}} \perp [1\,0\,0]_{\mathrm{bct}}$ for the dominant variant, mass density 8.12 g·cm^{-3}). At the room temperature, this sample was in tetragonal non-modulated (NM) martensite phase, consisting of a mixture of all three variants with one orientation being dominant. The sample was characterized with heating from the room temperature over the reverse transition temperature $A_S \approx 400$ K. In this case, the A_S temperature was above the Curie point (373 K), i.e., the transition occurred in the paramagnetic state. Similarly as in the case of sample No.2, an effective c' coefficient was used for characterizing the shear softening in non-modulated martensite below A_S; and,
4. a Ni-Ti sample (sample dimensions in austenite $3.48 \times 3.16 \times 2.78$ mm^3, sample orientation in austenite approximately $[1\,0\,0] \perp [0\,1\,0] \perp [0\,0\,1]$, mass density 6.50 g·cm^{-3}). This sample was characterized with cooling from 293 K towards the vicinity of the austenite→R-phase transition temperature $R_S \approx 270$ K. Twenty lowest resonant modes were traced in the measured temperature interval. Unlike samples Nos. 1, 2, and 3, the Ni-Ti sample was heterogeneous, consisting of a Ni-Ti matrix with finely dispersed precipitates. This means that the observed behavior (the elastic constants, the transformation temperatures, etc.) resulted from an interplay between the lattice (phonon mediated) instability of the matrix and the local stress fields that come from the precipitates. The discussion of such phenomena falls beyond the scope of this paper; here we will treat the measured elastic coefficients and their temperature dependencies as sufficiently representing the studied material at the macro-scale.

All of the samples were characterized using contact-less laser-based RUS arrangement described in detail in [21]. In this arrangement, the vibrations are generated by short focused laser pulses that act as broadband sources of acoustic waves; the detection is then done through a MSA-600 scanning laser vibrometer (Polytec GmbH, Waldbronn, Germany) that records the time-domain response of the sample in a pre-defined regular mesh covering the polished face of the sample. This scanning allows for the identification of modal shapes that correspond to the individual resonant peaks in the spectrum [32]. The resonant spectra were recorded in the frequency range 50 kHz–1 MHz, which typically covered up to 50 resonant modes. As only the softest shear coefficient c' was to be determined for most of the materials, only 10 or less lowest modes were taken for the inverse calculation; for the calculation itself, the Ritz–Rayleigh method-based procedure described in [21] was utilized. For the Ni-Ti sample (No.4), where the RUS data were also used for verifying the crystallographic orientation of the sample (see Section 3.3 for more details), an additional measurement in a frequency range up to 3 MHz was done. This enabled more than 120 resonant modes to be reliably identified and involved in the inverse calculation.

The internal friction parameters Q^{-1} were determined for the Cu-Al-Ni sample (No.1) from six dominant resonant peaks in each spectrum, as

$$Q^{-1} = \mathrm{FWHM}/f, \tag{4}$$

where f is the resonant frequency and FWHM stands for 'full width at half-maximum' for the given peak. For a reliable estimation of FWHMs, the peaks were fitted by a Lorenzian mask. For simplicity, we assumed that a single Q^{-1} parameter fully represents the viscoelastic behavior of the material. As all of the analyzed modes for the Cu-Al-Ni single crystal were the lowest modes corresponding only to the c'-shearing, this assumption was fully justified; this was also verified by the fact that the Q^{-1} parameters at given temperature were always very similar for all six analyzed modes. However, in more general cases, different straining modes of the material may correspond to different Q^{-1} parameters, and the anisotropy of the internal friction must be taken into account [25].

3. Results and Discussion

3.1. Cu-Al-Ni

For Cu-Al-Ni, the main aim was to analyze the behavior in the nearest vicinity of the transition temperature. The previous studies on this alloy [33–35] confirmed a gradual decrease of c' towards M_S, but it did not reveal any acceleration of the softening closer to the transition temperature. Also, the internal friction analysis reported by Sugimoto et al. [36] did not indicate any increase of the Q^{-1} parameter prior to the transition, while a steep increase in this parameter was observer below M_S. However, the loss of stability of the austenite lattice that triggers the martensitic transition should be visible as both an accelerated decrease of c' and an increase in the internal friction, both preceding the transition itself. This suggest that the experimental methods that were used in [33–36] did not have sufficient accuracy or sufficiently precise temperature control to capture these phenomena.

The laser-based RUS measurements can fulfill such a task very well, as shown in Figure 2. For the c'-softening (Figure 2a), a steep acceleration is observed below 302 K, i.e., just 3 K above the transformation temperature. This acceleration is accompanied by a rapid growth of the internal friction (Figure 2b), which increases by more than an order of magnitude.

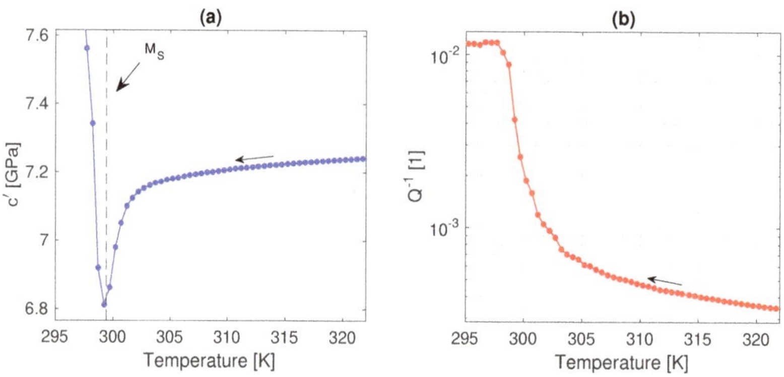

Figure 2. Evolution of the elastic constant c' (**a**) and the internal friction parameter (**b**) for the CuAlNi single crystal.

Below M_S, the material becomes a mixture of an austenitic matrix with nuclei of orthorhombic (2H) martensite. For small volume fractions of martensite, the mixture is still approximately cubic, but the effective c' for this mixture steeply increases due to the growth of the elastically stiffer martensite phase (see [37] for elastic constants of both phases). Surprisingly, the internal friction remained approximately constant, as it was probably mainly affected by the presence of the nuclei, not their volume fraction. Below 295 K, the material became so heterogeneous that the spectra were impossible to process. As the transition temperature interval for the Cu-Al-Ni alloy is relatively broad, $M_S - M_F \approx 15$ K [38,39], the RUS measurement between 295 K and $M_S = 299$ K covered only the very beginning of the formation of the thermally induced martensitic microstructure.

3.2. Ni-Mn-Ga

For the Ni-Mn-Ga alloy, the main aim was to compare the phonon-induced softening towards the transition temperature for paramagnetic and ferromagnetic austenite and martensite, and to discuss the effect of the ferromagnetic ordering on the softening. For this reason, two different alloys were measured (sample No.2 and sample No.3), one having the transition temperature below the ferromagnetic transition temperature (the Curie point) and one above it.

Figure 3a summarizes the results for sample No.2. For the austenite phase, a pronounced change of the dc'/dT slope occurs at the Curie point. As discussed in [40,41], this change is caused by the magneto-elastic coupling, which effectively softens the austenite lattice. In other words, below T_C the softening of the austenite phase can be decomposed into two parts: the first is ascribed to the TA_2 phonon-induced softening and retains approximately the same dc'/dT slope as above the Curie point (the dash-dot line in Figure 3a), and the second to the magneto-elastic coupling. As shown in [41], when the measurements are done in an external magnetic field, only the phonon-induced softening appears, and it accelerates very close to the transitions temperature in a similar manner as for Cu-Al-Ni discussed in the previous section (but the accelerated softening in Ni-Mn-Ga austenite that is reported in [41] is significantly stronger). However, here no such acceleration can be seen, most probably because such finer effects are completely overridden by the strong magneto-elastic softening.

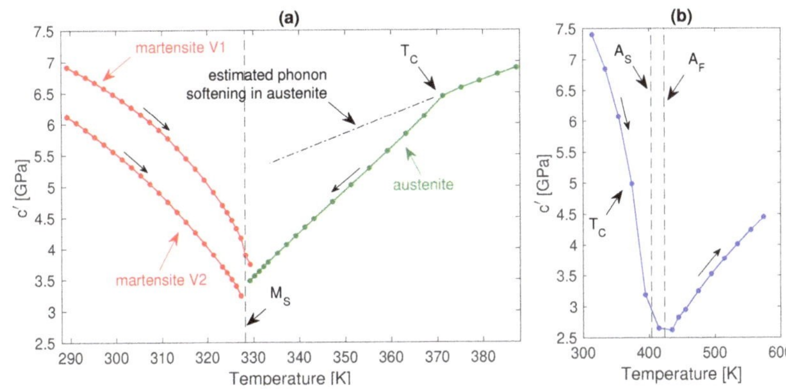

Figure 3. Evolution of the elastic constant c' for the Ni-Mn-Ga samples: (**a**) sample No.2 with the transition temperature below the Curie point T_C; (**b**) sample No.3 with the transition temperature above the curie point. In (**a**), the constants c' for two tetragonal variants of martensite V1 and V2 are effective elastic constants only, obtained under the assumption that the material has a cubic symmetry (see the text for more details); the dash-dot line is the extrapolation of the behavior of austenite above the Curie point. In (**b**), the c' below A_S is again an effective parameter, representing the initial mixture of variants. Note the different temperature scales between (**a**,**b**).

Below M_S, the measurements were done for two different tetragonal variants of martensite (in Figure 3a we use the denotation V1 for the variant having the $c-$axis oriented along the shortest edge of the sample and V2 for the variant having the $c-$axis oriented along the longest edge). Each of these variants had a slightly different effective c'; however, as seen in Figure 3a, the evolution of these c'-s with temperature was nearly identical. In both cases, a steep and further accelerated softening was observed towards the transition temperature. The variant V1 was slightly more stable, i.e., the reverse transition for it was started at a slightly higher temperature than for V2; nevertheless, the results confirmed that the temperature hysteresis for Ni-Mn-Ga is very small ($\sim$ 2 K) and the magnetically induced orientation did not lead to any significant mechanical stabilization of martensite (cf. [39]).

The measurement was repeated for a sample No.3 in order to observe the behavior of the soft shearing modes near the transition temperature without the magneto-elastic effects. This sample has a higher manganese content, which shifts the transition temperature above the Curie point. Moreover, this sample was in the non-modulated (NM) martensite at the room temperature, not in the 10 M modulated phase, due to the manganese excess. The Ni-Mn-Ga NM martensite cannot be reoriented by external magnetic field and, owing to its tetragonal distortion with $c/a > 1$, also the mechanical reorientation into a single variant would require loading the sample in tension [42]. As the tensile loading is nearly impossible for a sample of shape and dimensions suitable for RUS, the sample had to be measured in the as-received state consisting of a mixture of tetragonal variants. (Notice that the field-induced reorientation of NM martensite is not possible for pure Ni-Mn-Ga, but it can be achieved for Ni-Mn-Ga-based alloys doped by small amounts of Cu and Co [43], where the doping leads to changes in the soft shearing modes of the NM lattice [44].)

The sample was measured up to 573 K with a temperature step of 20 K (the longer step was chosen to avoid overly long exposure of the material to high temperatures, which might lead to undesired annealing processes). The result is seen in Figure 3b. Similarly as for the 10 M martensite, the NM martensite exhibits steep and accelerating softening with heating. The dc'/dT slope does not exhibit any sharp change at the Curie point, which is in good agreement with the fact that magneto-elastic coupling in NM martensite is negligible [45].

With further heating, the slope changes its sign approximately between 400 K and 420 K, which corresponds to the A_S and A_F temperatures, respectively. The temperature hysteresis for the NM martensite is much broader than for the 10 M martensite. This is probably the consequence of a larger strain misfit between austenite and NM martensite (i.e., λ_2 is much closer to 1 for the 10 M martensite, see [46,47] for the discussion of the relation between λ_2 and the hysteresis), and also of the much higher twinning stress in NM martensite.

In the austenite phase, which means above A_F, the shear softening is clearly visible, with some slight but visible acceleration towards the transition temperature. Interestingly, the c' that is close to the transition temperature reaches significantly lower values than for the sample No.2., where the softening includes both the phonon contribution and the contribution from the magneto-elastic coupling. This may mean that the paramagnetic austenite is even more prone to instability than the ferromagnetic austenite. Assuming that the c_{44} coefficient of Ni-Mn-Ga austenite only slightly decreases with temperature, reaching some $c_{44} \approx 100$ GPa at 420 K (estimated from the extrapolation of its room-temperature linear trend published in [48]), the anisotropy factor for sample No.3 at A_F can be estimated as $A \approx 38$, which is one of the highest values ever reported.

Finally, let us notice that the RUS analysis of the high-temperature spectra for this sample can be somehow affected by the uncertainty in the sample dimensions and the sample orientation in austenite. Both of these parameters were estimated from the dimensions and orientations of NM martensite, assuming the Bain path distortion with the tetragonal ratio $c/a = 1.21$ and approximate volume fractions of the minor variants. However, due to the strong anisotropy, the error in the orientation should not play any significant role (see the analysis in the Introduction), and the error due to the shape change should not exceed the level of the maximal possible error in the dimensions, which is $c/a - 1 = 0.21$ (an error obtained if the microstructure was completely different than that assumed). Thus, the real c' value at A_F falls into the range (2.6 ± 0.5) GPa, which is well below c' for sample No.2. and confirms the extremely strong anisotropy.

3.3. Ni-Ti

In the case of the Ni-Ti single crystal (sample No.4), a slightly different approach had to be applied. The main reason was that the transition mechanism in NiTi was not the Bain path distortion [4] and, thus, also the basal shear constant c_{44} was expected to be affected by the phonon softening [4,49–52]. In order to capture the behavior of both c' and c_{44}, a higher number of modes in RUS spectra had to be taken for the analysis. Furthermore, the anisotropy factor for NiTi austenite is relatively low ($A \approx 2$, [49]). As discussed in the introduction, this means that the determined value of c' might be affected by a possible misorientation of the sample. For this reason, the following strategy was adopted:

prior to the temperature-resolved measurement, the room temperature RUS data were thoroughly processed, assuming a general triclinic elastic anisotropy. The room-temperature RUS measurement was also complemented by TOF measurements in directions perpendicular to the individual faces of the sample. Using a combination of the RUS and TOF data, a full tensor including all 21 independent elastic coefficients was obtained, and the mirror-symmetry planes for this tensor were sought using the numerical procedure that was described in [21].

This analysis confirmed the cubic anisotropy of the sample, as nine distinct symmetry planes were found corresponding to all possible $\{1\,0\,0\}$ and $\{1\,1\,0\}$ planes. Additionally, a slight misorientation of the principal directions from the orientations of the edges was revealed, as shown in Figure 4a. The determined misorientaion was $\approx 3^{\circ}$, which is at the resolution limit of the Laue method. The spectra from the temperature measurements were then processed using this refined orientation of the sample, which resulted in the $c'(T)$ and $c_{44}(T)$ dependences plotted in Figure 4b. As expected based on the results from [49] (see also [50–52]), both of the constants exhibit softening towards the R-phase start (R_S) temperature. The softening of c_{44} is stronger and, unlike for c', exhibits slight acceleration towards the transition temperature. Nevertheless, no localized strong acceleration in the nearest vicinity of the transition temperature is measured (cf. Figure 2), although the temperature step above the transition temperature was relatively small (1 K). This can be due to the character of the R-phase transition, which is rather gradual and it has some second order-like features [53]. This behavior of the elastic constants is also in good agreement with the results for the aged $Ni_{51}Ti_{49}$ and $Ni_{50.8}Ti_{49}$, as reported in [49].

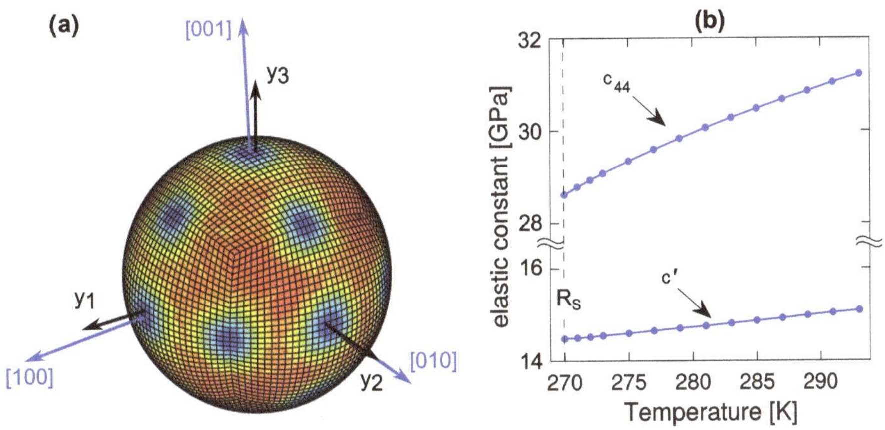

Figure 4. (**a**) the misorientation between the sample edges ($y_{1,...,3}$) and the principal $\langle 1\,0\,0 \rangle$ directions. The map plotted on the unit sphere is a map of a function used for determination of the symmetry planes of the material for the measured 21 (triclinic) elastic constants, as introduced in [21]. In particular, the sharp minima (blue spots) correspond to directions perpendicular to the mirror planes. (**b**) the resulting evolution of the elastic constants.

Thanks to the exactly known orientation of the sample, the elastic constants c' and c_{44} were obtained with very high accuracy. For all resonant modes involved in the inverse calculation, the experimental resonant frequencies differed from those that were computed from the resulting elastic constants by less then 0.25 kHz. Based on the sensitivity analysis described in [21], we can estimate that such a perfect match between the frequencies translates to a less than ± 0.1 GPa experimental uncertainty in c' and c_{44}. At the same time, however, the used RUS data did not allow for an accurate determination of c_{11} evolution with temperature. In other words, despite the higher number of modes and perfectly known orientation, the information that was carried by the RUS data was still dominantly on the shear elastic coefficients. Complementary temperature-resolved TOF measurements would be necessary for capturing the behavior of the whole cubic tensor.

4. Conclusions

Using three typical examples of SMAs (Cu-Al-Ni, Ni-Mn-Ga and Ni-Ti), we demonstrated that the laser-based RUS is a very suitable tool for the characterization of elastic precursors of martensitic transitions. The high accuracy and good temperature control allowed for a detailed examination of the behavior very close to the transition temperature, including the internal friction analysis. The access to high-temperature measurements due to the contactless character of the laser-based arrangement was shown to be beneficial for ferromagnetic SMAs, where the phonon-mediated softening had to be analyzed above the Curie point. Finally, we showed that the RUS data contained a sufficient information for a full characterization of an material with a moderate anisotropy factor. This was be initially utilized for improving the accuracy in the sample orientation, and then for the accurate determination of the evolutions of both c' and c_{44}.

Author Contributions: Conceptualization, H.S. and P.S.; investigation, M.J. and L.B.; resources, O.H.; data curation, M.J. and P.S.; writing–original draft preparation, H.S. and O.H.; writing–review and editing, O.H. All authors have read and agreed to the published version of the manuscript.

Funding: This work was supported by Czech Science Foundation, project No.20-12624S. The work of H. Seiner was supported by the Centre of Excellence for Nonlinear Dynamic Behaviour of Advanced Materials in Engineering CZ.02.1.01/0.0/0.0/15_003/0000493 (Excellent Research Teams) in the framework of Operational Programme Research, Development and Education. The work of O. Heczko was supported by the Czech Ministry of Education, Youth and Sports through Project MATFUN - CZ.02.1.01/0.0/0.0/15_003/0000487.

Conflicts of Interest: The authors declare no conflict of interest.

References

1. Delaey, L. Diffusionless transformations. In *Phase Transformations in Materials*; Kostorz, G., Ed.; Wiley-VCH: Weinheim, Germany, 2001; pp. 583–654.
2. Nakanishi, N. Elastic constants as they relate to lattice properties and martensite formation. *Prog. Mater. Sci.* **1980**, *24*, 143–265. [CrossRef]
3. Lüthi, B. *Physical Acoustics in the Solid State*; Springer: Berlin/Heidelberg, Germany, 2005.
4. Otsuka, K.; Ren, X. Physical metallurgy of Ti-Ni-based shape memory alloys. *Prog. Mater. Sci.* **2005**, *50*, 511–678. [CrossRef]
5. Chernenko, V.A.; Pons, J.; Seguí, C.; Cesari, E. Premartensitic phenomena and other phase transformations in Ni-Mn-Ga alloys studied by dynamical mechanical analysis and electron diffraction. *Acta Mater.* **2002**, *50*, 53–60. [CrossRef]
6. Muto, S.; Oshima, R.; Fujita, F.E. Elastic softening and elastic strain energy consideration in the f.c.c.-f.c.t. transformation of FePd alloys. *Acta Metall. Mater.* **1990**, *38*, 685–694. [CrossRef]
7. Stipcich, M.; Mañosa, L.; Planes, A.; Morin, M.; Zarestky, J.; Lograsso, T.; Stassis, C. Elastic constants of Ni-Mn-Ga magnetic shape memory alloys. *Phys. Rev. B Condens. Matter Mater. Phys.* **2004**, *70*, 054115. [CrossRef]
8. Worgull, J.; Petti, E.; Trivisonno, J. Behavior of the elastic properties near an intermediate phase transition in Ni_2MnGa. *Phys. Rev. B Condens. Matter Mater. Phys.* **1996**, *54*, 15695–15699. [CrossRef]
9. Musgrave, M.J.P. *Crystal Acoustics*; Holden Day: San Francisco, CA, USA, 1965.
10. Ledbetter, H.M. Sound velocities and elastic-constant averaging for polycrystalline copper. *J. Phys. D Appl. Phys.* **1980**, *13*, 1879–1884. [CrossRef]
11. Seiner, H.; Kopeček, J.; Sedlák, P.; Bodnárová, L.; Landa, M.; Sedmák, P.; Heczko, O. Microstructure, martensitic transformation and anomalies in c'-softening in Co-Ni-Al ferromagnetic shape memory alloys. *Acta Mater.* **2013**, *61*, 5869–5876. [CrossRef]
12. Zoubková, K. Elastic Response of Ferromagnetic Shape Memory Alloys in the Vicinity of the Critical Point. Master's Thesis, Czech Technical University, Prague, Czech Republic, 2018.

13. Papadakis, E.P.; Lerch, T.P. Pulse superposition, pulse-echo overlap and related techniques. In *Handbook of Elastic Properties of Solids, Liquids and Gases, 1–4*; Levy, M., Bass, H., Stern, R., Eds.; Academic Press: San Diego, CA, USA, 2001; pp. 39–66.
14. Migliori, A.; Sarrao, J.L.; Visscher, W.M.; Bell, T.M.; Lei, M.; Fisk, Z.; Leisure, R.G. Resonant ultrasound spectroscopic techniques for measurement of the elastic moduli of solids. *Phys. B Phys. Condens. Matter* **1993**, *183*, 1–24. [CrossRef]
15. Leisure, R.G.; Willis, F.A. Resonant ultrasound spectroscopy. *J. Phys. Condens. Matter* **1997**, *9*, 6001–6029. [CrossRef]
16. Hirao, M.; Ogi, H. *EMATs for Science and Industry: Noncontacting Ultrasonic Measurements*; Kluwer Academic Publishers: London, UK, 2003.
17. Vasil'ev, A.N.; Kokorin, V.V.; Savchenko, Y.I.; Chernenko, V.A. The magnetoelastic properties of a Ni_2MnGa single crystal. *Sov. Phys. JETP* **1990**, *71*, 803–805.
18. Landa, M.; Sedlák, P.; Seiner, H.; Heller, L.; Bicanová, L.; Šittner, P.; Novák, V. Modal resonant ultrasound spectroscopy for ferroelastics. *Appl. Phys. A Mater. Sci. Process.* **2009**, *96*, 557–567. [CrossRef]
19. Hurley, D.H.; Reese, S.J.; Park, S.K.; Utegulov, Z.; Kennedy, J.R.; Telschow, K.L. In situ laser-based resonant ultrasound measurements of microstructure mediated mechanical property evolution. *J. Appl. Phys.* **2010**, *107*, 063510. [CrossRef]
20. Wolfe, J. P. *Imaging Phonons (Acoustic Wave Propagation in Solids)*; Cambridge University Press: Cambridge, UK, 1998.
21. Sedlák, P.; Seiner, H.; Zídek, J.; Janovská, M.; Landa, M. Determination of All 21 Independent Elastic Coefficients of Generally Anisotropic Solids by Resonant Ultrasound Spectroscopy: Benchmark Examples. *Exp. Mech.* **2014**, *54*, 1073–1085. [CrossRef]
22. Sato, S.; Inakagi, K.; Gusev, V.E.; Wright, O.B. Resonant ultrasound spectroscopy using optical excitation and detection. *AIP Conf. Proc.* **1999**, *463*, 424–426.
23. Zadler, B.J.; Le Rousseau, J.H.L.; Scales, J.A.; Smith, M.L. Resonant ultrasound spectroscopy: Theory and application. *Geophys. J. Int.* **2004**, *156*, 154–169. [CrossRef]
24. Sumino, Y.; Ohno, I.; Goto, T.; Kumazawa, M. Measurement of elastic constants and internal frictions on single-crystal mgo by rectangular parallelepiped resonance. *J. Phys. Earth* **1976**, *24*, 263–273. [CrossRef]
25. Seiner, H.; Sedlák, P.; Koller, M.; Landa, M.; Ramírez, C.; Osendi, M.I.; Belmonte, M. Anisotropic elastic moduli and internal friction of graphene nanoplatelets/silicon nitride composites. *Compos. Sci. Technol.* **2013**, *75*, 93–97. [CrossRef]
26. Straka, L.; Drahokoupil, J.; Pacherová, O.; Fabiánová, K.; Kopecký, V.; Seiner, H.; Hänninen, H.; Heczko, O. The relation between lattice parameters and very low twinning stress in Ni50Mn25+xGa25-x magnetic shape memory alloys. *Smart Mater. Struct.* **2016**, *25*, 025001. [CrossRef]
27. Sozinov, A.; Likhachev, A.A.; Lanska, N.; Söderberg, O.; Ullakko, K.; Lindroos, V.K. Stress- and magnetic-field-induced variant rearrangement in Ni-Mn-Ga single crystals with seven-layered martensitic structure. *Mater. Sci. Eng. A* **2004**, *378*, 399–402. [CrossRef]
28. Scheerbaum, N.; Heczko, O.; Liu, J.; Hinz, D.; Schultz, L.; Gutfleisch, O. Magnetic field-induced twin boundary motion in polycrystalline Ni-Mn-Ga fibres. *New J. Phys.* **2008**, *10*, 073002. [CrossRef]
29. Straka, L.; Heczko, O.; Seiner, H.; Lanska, N.; Drahokoupil, J.; Soroka, A.; Fähler, S.; Hänninen, H.; Sozinov, A. Highly mobile twinned interface in 10 M modulated Ni-Mn-Ga martensite: Analysis beyond the tetragonal approximation of lattice. *Acta Mater.* **2011**, *59*, 7450–7463. [CrossRef]
30. Zhao, P.; Dai, L.; Cullen, J.; Wuttig, M. Magnetic and elastic properties of Ni49.0Mn23.5 Ga27.5 premartensite. *Metall. Mater. Trans. A Phys. Metall. Mater. Sci.* **2007**, *38*, 745–751. [CrossRef]
31. Dai, L.; Cui, J.; Wuttig, M. Elasticity of Austenitic and Martensitic NiMnGa. *Proc. SPIE Int. Soc. Opt. Eng.* **2003**, *5053*, 595–602.
32. Ogi, H.; Sato, K.; Asada, T.; Hirao, M. Complete mode identification for resonance ultrasound spectroscopy. *J. Acoust. Soc. Am.* **2002**, *112*, 2553–2557. [CrossRef]
33. Recarte, V.; Pé érez-Landazábal, J.I.; Nó, M.L.; San Juan, J. Study by resonant ultrasound spectroscopy of the elastic constants of the β phase in Cu-Al-Ni shape memory alloys. *Mater. Sci. Eng. A* **2004**, *370*, 488–491. [CrossRef]

34. Landa, M.; Sedlák, P.; Šittner, P.; Seiner, H.; Heller, L. On the evaluation of temperature dependence of elastic constants of martensitic phases in shape memory alloys from resonant ultrasound spectroscopy studies. *Mater. Sci. Eng. A* **2008**, *481–482*, 567–573. [CrossRef]
35. Landa, M.; Sedlák, P.; Šittner, P.; Seiner, H.; Novák, V. Temperature dependence of elastic properties of cubic and orthorhombic phases in Cu-Al-Ni shape memory alloy near their stability limits. *Mater. Sci. Eng. A* **2007**, *462*, 320–324. [CrossRef]
36. Sugimoto, K.; Mori, T.; Otsuka, K.; Shimizu, K. Simultaneous measurements of internal friction, young's modulus and shape change associated with thermoelastic martensite transformation in CuAlNi single crystals. *Scr. Metall.* **1974**, *8*, 1341–1348. [CrossRef]
37. Sedlák, P.; Seiner, H.; Landa, M.; Novák, V.; Šittner, P.; Mañosa, L. Elastic constants of bcc austenite and 2H orthorhombic martensite in CuAlNi shape memory alloy. *Acta Mater.* **2005**, *53*, 3643–3661. [CrossRef]
38. Seiner, H.; Sedlak, P.; Landa, M. Shape recovery mechanism observed in single crystals of Cu-Al-Ni shape memory alloy. *Phase Transitions* **2008**, *81*, 537–551. [CrossRef]
39. Vronka, M.; Seiner, H.; Heczko, O. Temperature dependence of twinning stress–Analogy between Cu–Ni–Al and Ni–Mn–Ga shape memory single crystals. *Philos. Mag.* **2017**, *97*, 1479–1497. [CrossRef]
40. Seiner, H.; Heczko, O.; Sedlák, P.; Bodnárová, L.; Novotný, M.; Kopeček, J.; Landa, M. Combined effect of structural softening and magneto-elastic coupling on elastic coefficients of NiMnGa austenite. *J. Alloy. Compd.* **2013**, *577*, S131-S135. [CrossRef]
41. Seiner, H.; Sedlák, P.; Bodnárová, L.; Drahokoupil, J.; Kopecký, V.; Kopeček, J.; Landa, M.; Heczko, O. The effect of antiphase boundaries on the elastic properties of Ni-Mn-Ga austenite and premartensite. *J. Phys. Condens. Matter* **2013**, *25*, 425402. [CrossRef] [PubMed]
42. Söderberg, O.; Straka, L.; Novák, V.; Heczko, O.; Hannula, S.-P.; Lindroos, V.K. Tensile/compressive behaviour of non-layered tetragonal Ni52.8Mn25.7Ga21.5 alloy. *Mater. Sci. Eng. A* **2004**, *386*, 27–33. [CrossRef]
43. Sozinov, A.; Lanska, N.; Soroka, A.; Zou, W. 12% magnetic field-induced strain in Ni-Mn-Ga-based non-modulated martensite. *Appl. Phys. Lett.* **2013**, *102*, 021902. [CrossRef]
44. Bodnárová, L.; Zelený, M.; Sedlák, P.; Straka, L.; Heczko, O.; Sozinov, A.; Seiner, H. Switching the soft shearing mode orientation in Ni-Mn-Ga non-modulated martensite by Co and Cu doping. *Smart Mater. Struct.* **2020**, *29*, 045022. [CrossRef]
45. Heczko, O.; Seiner, H.; Sedlák, P.; Kopeček, J.; Kopecký, V.; Landa, M. Resonant ultrasound spectroscopy—A tool to probe magneto-elastic properties of ferromagnetic shape memory alloys. *Eur. Phys. J. B* **2013**, *86*, 62. [CrossRef]
46. James, R.D.; Zhang, Z. A Way to Search for Multiferroic Materials with "Unlikely" Combinations of Physical Properties. *Springer Ser. Mater. Sci.* **2005**, *79*, 159–175.
47. Cui, J.; Chu, Y.S.; Famodu, O.O.; Furuya, Y.; Hattrick-Simpers, J.; James, R.D.; Ludwig, A.; Thienhaus, S.; Wuttig, M.; Zhang, Z.; et al. Combinatorial search of thermoelastic shape-memory alloys with extremely small hysteresis width. *Nat. Mater.* **2006**, *5*, 286–290. [CrossRef]
48. Seiner, H.; Kopecký, V.; Landa, M.; Heczko, O. Elasticity and magnetism of Ni_2MnGa premartensitic tweed. *Phys. Status Solidi (B) Basic Res.* **2014**, *251*, 2097–2103. [CrossRef]
49. Ren, X.; Miura, N.; Zhang, J.; Otsuka, K.; Tanaka, K.; Koiwa, M.; Suzuki, T.; Chumlyakov, Y.I.; Asai, M. A comparative study of elastic constants of Ti-Ni based alloys prior to martensitic transformation. *Mater. Sci. Eng. A* **2001**, *312*, 196–206. [CrossRef]
50. Brill, T.M.; Mittelbach, S.; Assmus, W.; Mullner, M.; Luthi, B. Elastic properties of NiTi. *J. Phys. Condens. Matter* **1991**, *3*, 9621–9627. [CrossRef]
51. Mercier, O.; Melton, K.N.; Gremaud, G.; Hägi, J. Single-crystal elastic constants of the equiatomic NiTi alloy near the martensitic transformation. *J. Appl. Phys.* **1980**, *51*, 1833–1834. [CrossRef]
52. Ren, X.; Otsuka, K. The role of softening in elastic constant c_{44} in martensitic transformation. *Scr. Mater.* **1998**, *38*, 1669–1675. [CrossRef]

53. Šittner, P.; Landa, M.; Lukáš, P.; Novák, V. R-phase transformation phenomena in thermomechanically loaded NiTi polycrystals. *Mech. Mater.* **2006**, *38*, 475–492. [CrossRef]

Publisher's Note: MDPI stays neutral with regard to jurisdictional claims in published maps and institutional affiliations.

metals

MDPI

Article

{111}<110> Orientation Induced Anisotropy of Shape Memory Effect in NiTiNb Pipe Joints

Mingyan Sun, Qichao Fan, Yingying Wang, Qin Yang, Jie Chen, Shuke Huang * and Yonghao Zhang *

Institute of Machinery Manufacturing Technology, China Academy of Engineering Physics, Mianyang 621900, China; yanmingsun18@caep.cn (M.S.); fanqichao@caep.cn (Q.F.); yingying89@caep.cn (Y.W.); yangqin_6s@caep.cn (Q.Y.); jiechen_6s@caep.cn (J.C.)

* Correspondence: huangshuke@caep.cn (S.H.); zyhao_6s@caep.cn (Y.Z.); Tel.: +86-0816-2487607 (S.H.); +86-0816-2490543 (Y.Z.)

Received: 28 April 2020; Accepted: 4 June 2020; Published: 11 June 2020

Abstract: This work aims to clarify the influence of texture type and intensity on the shape memory effect (SME) in NiTiNb shape memory alloy (SMA) pipe joints, especially revealing the causes for the anisotropy of SME via texture changes. Three NiTiNb rods with different intensities of the {111}<110> texture were fabricated, and their microstructures, crystalline orientation distribution functions and inverse pole figures were obtained by X-ray diffraction and electron backscatter diffraction measurements. Simultaneously, the SME was characterized by inner-diameter recoverability of the corresponding pipe joints. For a given intensity of the {111}<110> texture, the SME of the NiTiNb pipe joints strongly depended on the expansion direction due to {111}<110> orientation-induced anisotropy of SME. In addition, both the SME and anisotropy of NiTiNb pipe joints increased with the increased intensity of the {111}<110> texture. Therefore, a suitable expansion direction and strong texture intensity should be considered for high SME in NiTiNb pipe joints.

Keywords: NiTiNb; anisotropy; texture; SME; pipe joints

1. Introduction

NiTi-based shape memory alloys (SMAs), due to their unique shape memory effect (SME), have been wildly used in many fields such as aerospace, biomedicine, mechanical electronics and automotive industries [1–4]. Among them, one of the most important and successful applications is the pipe joint [5–7]. NiTiNb alloys, especially those with a nominal composition of $Ni_{47}Ti_{44}Nb_9$ (at%), have attracted much attention as SMAs because they demonstrate wide transformation hysteresis after pre-deformation [8–12]. Thus, pipe joints constructed from NiTiNb alloys do not require storing and installing at low temperatures, which is quite useful for the engineering uses [13–15].

Generally, pipe joints are machined from NiTiNb rods and then expanded at low temperature to augment their inner diameter [16,17]. According to previous studies, in order to store and ship NiTiNb pipe joints at room temperature and simultaneously optimize the SME, researchers mainly pay attention to the key problem of finding appropriate expansion temperatures and critical expansion strains [18–22]. However, the expansion direction is rarely considered, despite the anisotropy of SME induced by texture in NiTiNb rods. Identifying the expansion direction that optimizes the SME is therefore crucial for fabricating high-performance NiTiNb pipe joints. The expansion direction can be controlled by preparing the pipe joints along the radial or axial direction of the rods.

As is well known, the NiTiNb rods used for pipe joints are typically prepared by vacuum melting, followed by thermomechanical deformation and finally by a suitable heat-treatment [23–26]. These processes inevitably generate crystallographic textures that profoundly affect the SME. It is necessary to understand which type of orientation features a higher reversibility and the extent to which texture intensity should be improved to satisfy pipe joints' application. Currently, studies on texture of NiTiNb are mainly focused on the effect and control of deformation and heat-treatment on the final texture types [26–29]. For example, Yan et al. found that the main texture types in $Ni_{47}Ti_{44}Nb_9$ hot-rolled rods concentrate on γ fiber (<111> // axial direction), while heat-treatment only decreases its intensity [26]. In addition, Yin et al. obtained <113> fiber texture in $Ni_{47}Ti_{44}Nb_9$ hot-forged rods, which turned into a strong γ fiber texture after cold-drawing [27]. In contrast, the relationship between texture and SME has been rarely investigated. The few studies on this topic have concentrated on the variations of recovery strain between the rolling direction (RD) and transverse direction (TD) in thin sheets, whereas pipe joints are generally prepared from rods [26]. Moreover, the positive role of texture in the SME of NiTiNb pipe joints has not been clearly clarified, because the expansion stress states of these joints are far more complex than those of uniaxial tensile or compressed specimens.

In this study, using $Ni_{47}Ti_{44}Nb_9$ as an example, three rods with different intensities of the {111}<110> texture were investigated. The study focused on clarifying the influence of texture type and intensity on the SME of $Ni_{47}Ti_{44}Nb_9$ pipe joints, especially revealing the causes for the anisotropy of SME via texture changes. The aim is to provide referential data for engineering applications.

2. Experimental Procedures

As shown in Figure 1, first, vacuum induction melting was used to produce the master ingot of $Ni_{47}Ti_{44}Nb_9$ (at%) with a diameter of 150 mm and a height of 200 mm. The master ingot was multi-directionally forged at 900 °C into 35 mm diameter rods for the sake of improving and homogenizing the microstructure of the as-cast ingot. To account for the influence of texture intensity on the resulting SME, the rods were rotary hot-forged along the axial direction by different passes to diameters of 15 mm, 10 mm and 8 mm (forming Samples A, B and C, respectively). Then, all rods were heat-treated in an evacuated quartz tube at 900 °C for 2 h followed by air quenching (AC) from high temperature to room temperature without cryogenic treatment. Finally, all samples were electric-discharge machined into the sizes needed for relevant measurements.

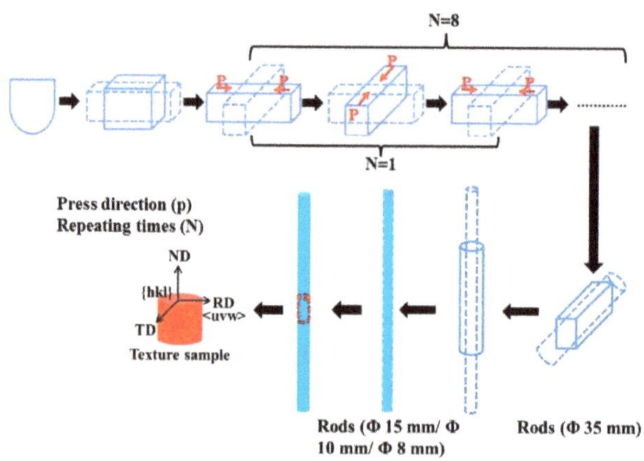

Figure 1. Schematic of the fabrication process of hot-forged rods and the orientations of the samples cut from the rods for texture measurements.

Phase identification and texture measurements were carried out by Cu K_α radiation using X-ray diffractometer (Empyrean, Panalytical, Almelo, Netherlands). The X-ray diffraction (XRD) patterns of the three samples are similar, as shown in Figure 2. The pole figures from the crystallographic planes {110}, {200}, {211} of the B_2 phase were measured at α = 0–70° and β = 0–360° with a step size of 5° and presented as orientation distribution function (ODF) charts. The texture-test Samples A, B and C (with dimensions of 15 mm, 10 mm and 8 mm in diameter, respectively, and 5 mm in length) were machined in the excision direction, which is perpendicular to the axial direction of the rods, as seen in Figure 1. Thus, the measured texture component {hkl}<uvw> means that the {hkl} planes perpendicular to the axial direction of the rods, the <uvw> directions aligned along the radial direction and the <rst> directions aligned along the circumferential direction.

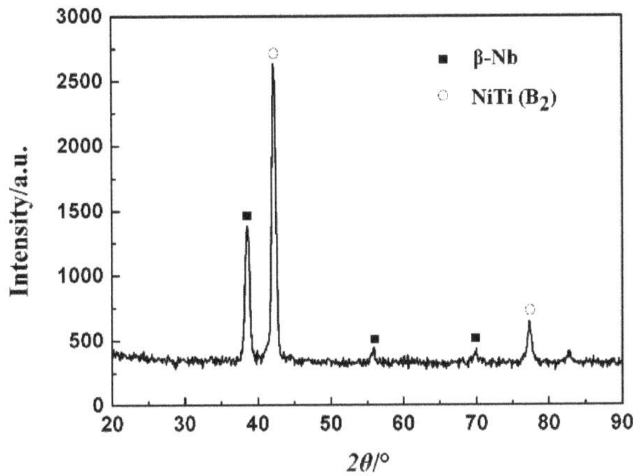

Figure 2. X-ray diffraction (XRD) patterns of Samples A, B and C, with the β-Nb and NiTi (B_2) peaks identified.

The phase transformation behavior was determined by differential scanning calorimeter (DSC, DSC 214 Polyma, Netzsch, Selb, Germany) at a 10 °C/min heating and cooling rate. DSC specimens (with dimensions of 1 × 1 × 2 mm^3) were chemically washed in a mixed acid solution.

The microstructures were characterized by electron backscatter diffraction (EBSD) to reveal their crystallographic grain boundaries and grain orientations. Both cross sections and longitudinal sections of the rods were tested. Specimens for EBSD were mechanically polished and then vibratory polished in colloidal silica. Field emission scanning electron microscope (SEM, JSM-6700F, JEOL Ltd., Tokyo, Japan) equipped with an EBSD detector and data analysis software (OIM^{TM}, TSL-EDAX, Mahwah, NJ, USA) was used for EBSD test. For a convenient analysis, the normal direction (ND) of the EBSD test zones was set perpendicular to the test plane of the rods and the RD for specimens with cross section and longitudinal section paralleled the radial direction and axial direction of the rods, respectively.

Tensile tests were carried out in an electronic universal testing system equipped with an environmental experiment box under a loading speed of 0.5 mm/min. Dog-bone-shaped tensile specimens were cut to a gauge length of 7 mm and a width of 1 mm, along the axial and radial direction of the rods, as shown in Figure 3a.

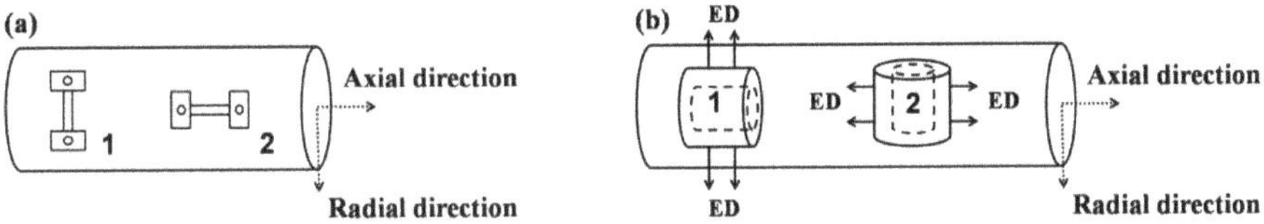

Figure 3. Schematic of the samples used for (**a**) the tensile test and (**b**) the shape memory effect (SME) test.

To ensure that the results closely matched those of actual engineering applications, the SME was characterized by inner-diameter recoverability of real pipe joints. The pipe joints were machined with an initial inner diameter (D_0), a wall thickness of 1.5 mm and the expansion direction (ED) along the radial and axial direction of the rods, as shown in Figure 3b. Then, the pipe joints were heat-treated at 900 °C for 2 h in an evacuated quartz tube and then expanded at −60 °C by using a core bar. After the inner diameter (D_1) was measured, the pipe joints were heated to 200 °C for 10 min to complete the inverse martensitic transformation. Then the inner diameter was again measured and recorded as D_2 The recovery rate (η) and recovery strain (ε_r) of the pipe joints were calculated according to the following formulas:

$$\eta = \frac{D_1 - D_2}{D_1 - D_0} \times 100\% \quad (1)$$

$$\varepsilon_r = \frac{D_1 - D_2}{D_0} \times 100\% \quad (2)$$

3. Results

3.1. Texture Comparison and EBSD Characterization

The ODF results of Samples A, B and C are shown in Figure 4. The preferred orientation of all samples concentrates on {111}<110>, as shown in the $\varphi2$ = 45° sections of Figure 4a–c. According to Yan's analysis, on rotary hot forging, the radial direction of rod is subjected to compressive stress, slip planes rotate toward the direction perpendicular to the external stress axis and slip directions toward the plastic flow direction or to the rod axis [23]. Thereby, the {111}<110> texture formed in the hot-forged rods are mainly related to the activation of {111}<110> slip systems in the B_2 phase. Meanwhile, the maximum orientation density of {111}<110>, from low to high, is ordered as A, B and C, indicating that the texture intensity can be controlled by the deformation degree. The inverse pole figure (IPF) maps of above three samples are obtained by EBSD test to present grain orientation (Figure 5). A strong {111}<110> texture is also observed in all samples, with grains in blue (<111> // axial direction) crossing most of the cross section and grains in green (<110> // radial direction) crossing most of the longitudinal section of the rods. This is consistent with the result tested by XRD above. In addition, grains of different shape with equiaxed grains on cross section (Figure 5a–c) and elongated grains on longitudinal section (Figure 5d–f) are observed, which means that the grains are elongated along the axial direction. Meanwhile, with the increase of deformation degree, the average grain size decreases and the grain morphology is further elongated.

3.2. Tensile Tests and Corresponding Anisotropic SME in the Rods

In order to clearly investigate the {111}<110> orientation-induced anisotropic SME in $Ni_{47}Ti_{44}Nb_9$ pipe joints, the deformation mode of uniaxial tension in $Ni_{47}Ti_{44}Nb_9$ rods is considered firstly. Taking Sample A as an example, dog-bone shaped tensile specimens are loaded at temperature of −60 °C, along the axial and radial direction (that is <111> and <110> direction, respectively) of the rods. As shown in Figure 6, the anisotropic stress–strain curves exhibit different length of phase transformation plateaus and different plateau stresses along the <111> and <110> directions. Specifically, the phase transformation plateau is longer along the <111> direction than along <110> direction and the plateau stress is higher along the <111> direction than that along the <110> direction. These results suggest that the stress-induced martensitic critical stress σ_{sim} along the <111> direction is larger. Meanwhile, the strain of the martensite nominal yield point (the nominal starting point that the dislocations begin to slip in martensite) is obviously larger along the <111> direction than along the <110> direction. In addition. after tension to 16%, the specimens are heated at 200 °C to calculate their recovery property. The recovery strain along the <111> direction is 9.7%, versus 9.1% along the <110> direction.

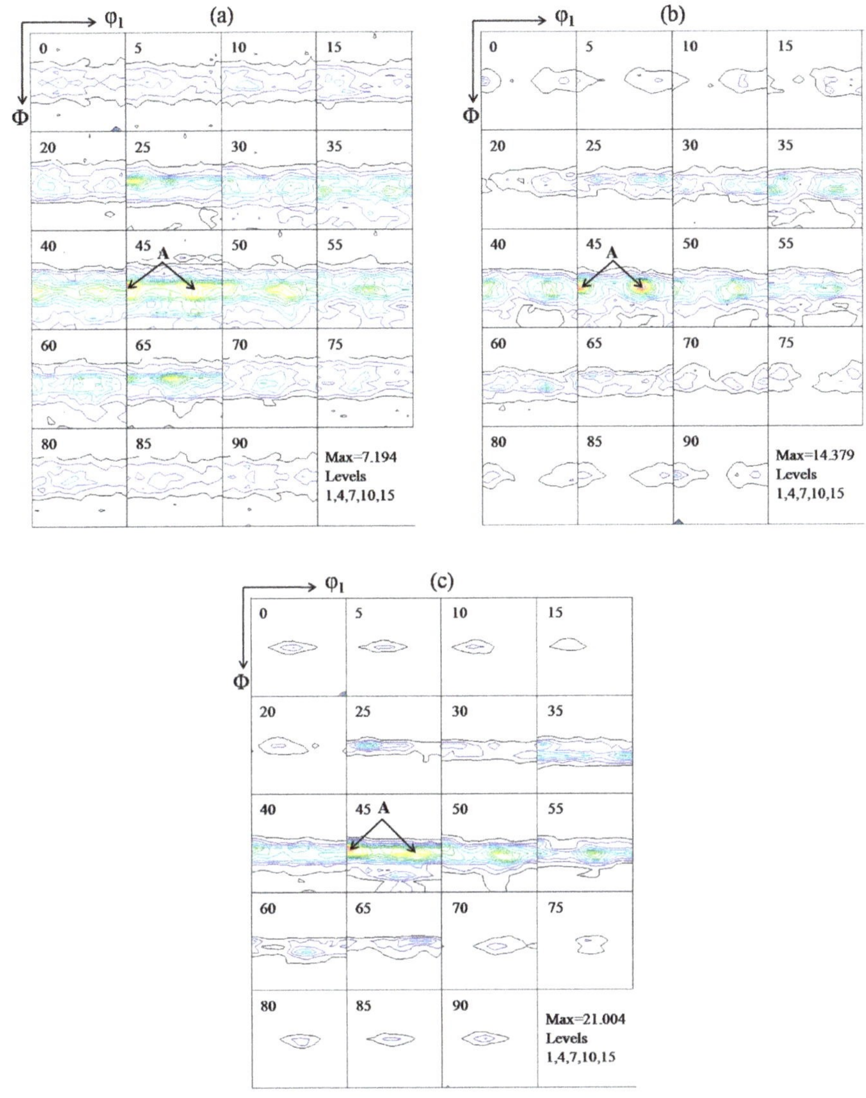

Figure 4. Orientation distribution function (ODF) results of the $Ni_{47}Ti_{44}Nb_9$ rods. (**a**) Sample A, (**b**) Sample B and (**c**) Sample C. Point A represents the {111}<110> component.

3.3. *Recoverability of the Pipe Joins*

Six pipe joins in each sample were measured for SME. The average value of η and ε_r are shown in Figure 7a,b, respectively. For the same expanding direction of ED1 or ED2, the average value of η and ε_r from low to high, is ordered as A, B and C. Meanwhile, in each sample, both η and ε_r along ED1 are higher than along ED2. In addition, the difference of recoverability between ED1 and ED2 increases on the order of A, B and C. These results reflect that Sample C with strongest texture of {111}<110> has the highest recoverability and the strongest anisotropy of recoverability and the reasons are discussed in the following section.

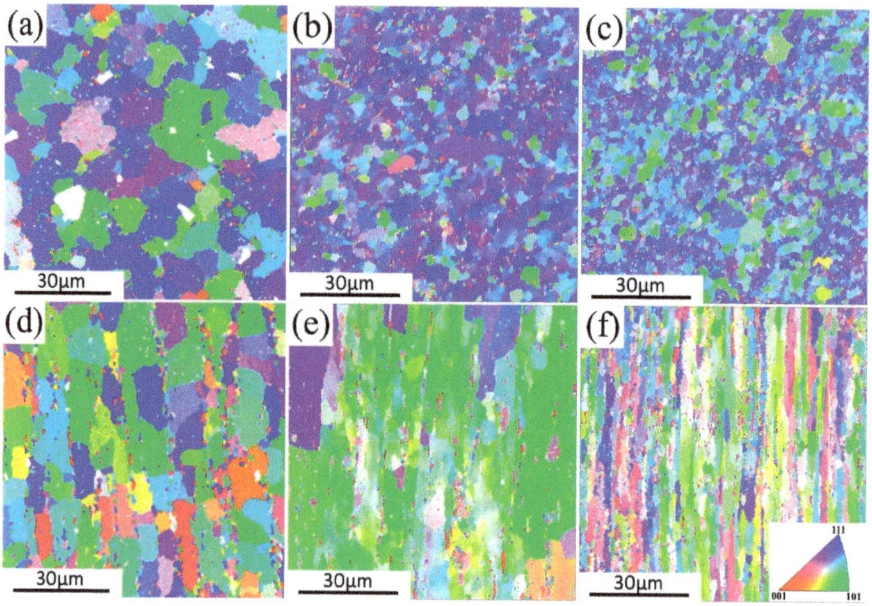

Figure 5. Inverse pole figure (IPF) maps of the NiTi (B_2) phase in different states, representing the preferred crystalline orientation in the normal direction (ND) of the test planes. (**a–c**) cross sections of Samples A, B and C, respectively; (**d–f**) longitudinal sections of Samples A, B and C, respectively.

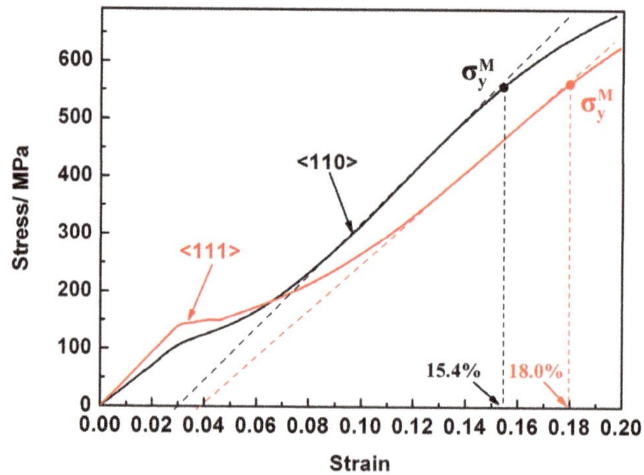

Figure 6. Tensile curves obtained during loading at −60 °C along the <111> and <110> direction. The martensite yield nominal point is got from the intersection of true stress–strain curve with the martensite elastic stage tangent after 0.2% horizontal movement.

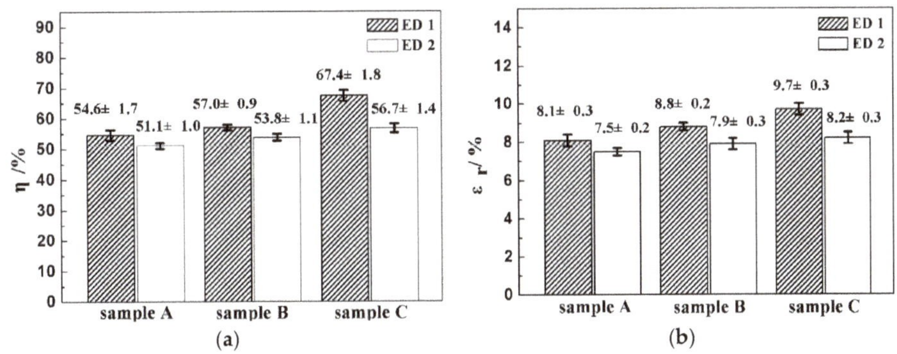

Figure 7. Recoverability of Samples A, B and C along the expansion directions of ED1 and ED2: (a) recovery rate η and (b) recovery strain ε_r.

4. Discussions

4.1. Effects of {111}<110> Orientation on SME in the Uniaxial Tensile Samples

As is well known, the SME of SMA is determined by the crystallographic reversibility of the reverse martensite transformation. Therefore, different lattice orientations lead to distinct recoverability. Previously, several researchers concentrating on NiTi single crystals have shown that the tensile recovery strains of <001>, <110>, <111> are 2.7%, 8.4%, 9.8%, respectively, while compressive recovery strains of <001>, <110>, <111> are 4.2%, 5.2%, 3.6%, respectively [30,31]. As for $Ni_{47}Ti_{44}Nb_9$ polycrystals, it is NiTi phase which plays the main role in recoverability, thus the recovery strain of the polycrystalline alloy can be computed as:

$$\overline{\varepsilon}_M = \sum_{i=1}^{n} \varepsilon_{ri} I_i \quad (3)$$

Here $\overline{\varepsilon}_M$ is the average recovery strain, ε_{ri} is the recovery strain in each orientation, I_i is the proportion of each orientation among the total orientations, and n is the number of the orientations. Therefore, as for the uniaxial tensile samples, the recovery strain along the <111> direction is larger than that along the <110> direction, which is in agreement with above experimental results.

To deeply understand the causes for anisotropic SME between the <111> and <110> direction, the tensile curves are analyzed in detail. According to previous studies on NiTiNb, the phase transformation plateau in stress–strain curves is formed by stress-induced martensite transformation and reorientation, the reversibility of which contributes to the strain recovery [32,33]. Therefore, when subject to the same deformation strain, a long plateau generally indicates a large recovery strain. As shown in Figure 6, the plateau is longer along the <111> direction than along the <110> direction, so the most favorably oriented martensite variants originating from <111> can generate larger strain than that from <110>, thus resulting in high SME. In addition, the strain of martensite nominal yield point along <111> direction is obviously larger than that along <110> direction. Thus, when loaded to the same strain of 16%, dislocations are easily generated along the <110> direction, which will partly impede the reverse transformation, according to the previous research that deformation-induced dislocations/vacancies are considered to be related to the martensite stabilization [34]. Therefore, from this point of view, the recovery strain along <111> direction is also larger than that along <110> direction.

For more details, Figure 8 shows the DSC curves before and after 16% tension along the <111> and <110> directions at −60 °C. Shown in figure, the A_s and A_f are largely increased after tension. In addition, the reverse transformation temperature of A_s' along the <111> direction is 69 °C, which is lower than 72 °C along the <110> direction, indicating that the reverse transformation occurs more easily along the <111> direction. The reason for the difference in the transformation temperatures of both conditions presented is related to the martensite stabilization introduced by deform-induced dislocations/vacancies.

Samples with more dislocations tend to highly impede the reverse transformation and make the reverse transformation temperature higher. The result also provides evidence that <111> direction tends to generate fewer dislocations, thus achieve higher recoverability, as is mentioned above.

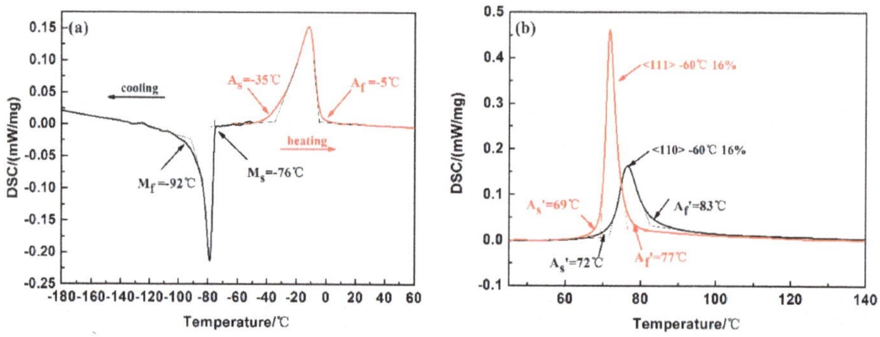

Figure 8. Differential scanning calorimetry (DSC) curves (**a**) before and (**b**) after tension to 16% at −60 °C along the <111> and <110> direction.

4.2. Effects of {111}<110> Orientation on SME in Pipe Joints

Different from uniaxial tensile sample, as for pipe joints, it should be analyzed in cylindrical coordinate rather than rectangular coordinate. With the inner diameter D_0 expands to D_1 in the cross section, the mechanical strain ε can be divided into a radial compressive strain ε_C and a circumferential tensile strain ε_L, as shown in Figure 9. During the heating process, the recovery of ε_C and ε_L make the inner diameter D_1 decreasing to D_2, thus realizing the connection of the two pipes. According to previous studies, the SME of pipe joints is mainly determined by circumferential strain ε_L rather than radial strain ε_C [27]. Hence, the recovery strain of pipe joints can be calculated as

$$\overline{\varepsilon}_M = \sum\nolimits_{i=1}^{n} \varepsilon_{Lri} I_i \quad (4)$$

Here ε_{Lri} is the recovery strain originating from the circumferential tensile strain ε_L in each orientation.

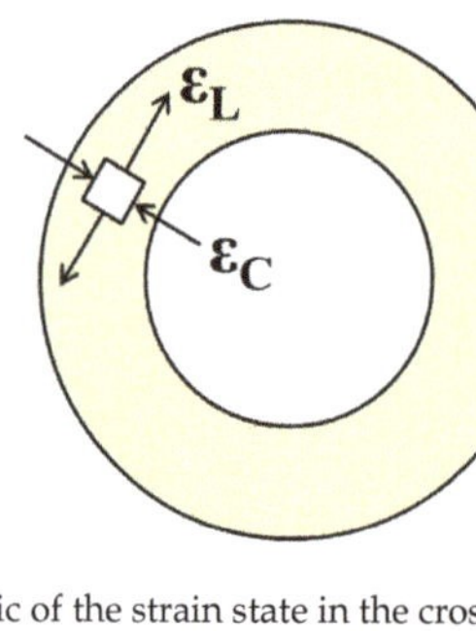

Figure 9. Schematic of the strain state in the cross section of a pipe joint.

Therefore, in samples of A, B, C with the same ED, the texture types are the same, while the texture intensity is different. In the order of A, B and C, the intensity of {111}<110> texture increases gradually, so the recoverability increases too, which is consistent with the results obtained in Figure 7.

In the same sample with different expansion directions of ED1 and ED2, the texture intensity is fixed, so the $\overline{\varepsilon}_M$ is mainly determined by the ε_{Lri} of the preferred orientation along the circumferential direction of the pipe joints. Taking sample C for an example, the IPF of ND, RD and TD is shown in Figure 10. There is high density preferred orientation in <111> district of ND IPF and <110> district of RD IPF and TD IPF, indicating that the <111> direction is parallel to the axial direction while the <110> direction is parallel to the radial and circumferential directions of the rod. Hence, when the pipe joint is expanded along ED1, the preferred orientation is <110> and this preferred orientation causes higher recovery strain, as is performed in single crystal materials. However, when expanded along ED2, the preferred orientation of circumferential direction in pipe joint is quite complicated, maybe including many orientations such as <111>, <110> and <001> with different recovery strains, thus the average recovery strain along ED2 is supposed to be smaller than that along ED1 due to the imbalance of recovery strain in many orientations.

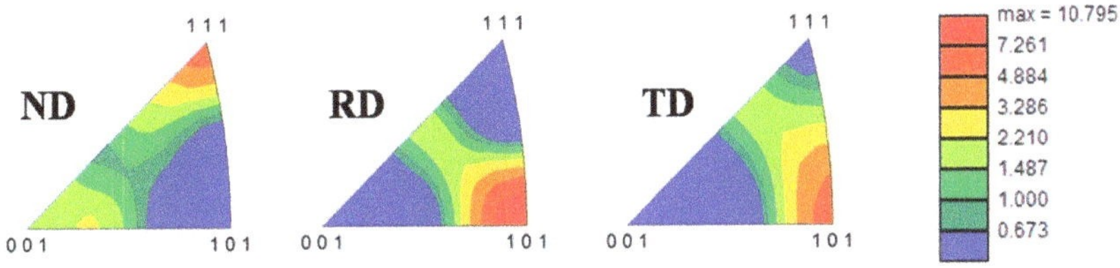

Figure 10. Inverse pole figures (IPFs) of Sample C, representing the preferred crystalline orientations in the axial direction (ND), radial direction (RD) and circumferential direction (TD) of the rod.

In addition. the reason that the recoverability difference between ED1 and ED2 increases on the order of A, B and C can be related to the texture intensity obviously. The stronger texture intensity the rod has, the bigger the recoverability difference is. Hence, strong texture intensity can strengthen the anisotropy of the SME in $Ni_{47}Ti_{44}Nb_9$ rods.

It is worth noting that, along with texture, grain size and grain morphology may also contribute to anisotropic SME. Both slender and small grains are expected to increase the anisotropy of SME. This trend may be explained by grain-boundary strengthening theory. Samples with more grain boundaries tend to own higher martensite yield stress. Thus, when loaded to the same strain, samples with more grain boundaries tend to generate fewer dislocations and exhibit better recoverability. However, it is still hard to know which factor plays a dominant role, since grain size, grain morphology and texture usually change together under deformation, and it is very hard to study a single factor without changing others. Obviously, to further clarify the cause, more work needs to be done.

5. Conclusions

1. The uniaxial tensile recovery strain along the <111> direction was larger than that along the <110> direction in NiTiNb rods;

2. For the same texture type of {111}<110> with the same expansion direction, the ε_r and η of the NiTiNb pipe joints increase along with increasing texture intensity. Thus, a strong texture intensity is desired in engineering applications;

3. For the same texture type of {111}<110> with the same texture intensity, the ε_r and η of the pipe joints along ED1 were higher than along ED2, indicating that the recoverability of NiTiNb pipe joints strongly depends on the expansion direction. Thus, the suitable expansion direction should be selected to improve the SME in pipe joints;

4. The recoverability difference between ED1 and ED2 increased along with increasing texture intensity, suggesting that a strong texture intensity further strengthens the anisotropy of the SME in NiTiNb rods.

Author Contributions: M.S. and S.H. proposed the main idea; M.S. and Q.F. made investigation; M.S. and Y.W. designed experiments with the help of Q.F.; M.S., Q.F. and Y.W. collected and analysed the data with the help of Q.Y., J.C. and Y.Z., M.S. wrote the original draft; Q.F., S.H. and Y.Z. reviewed and edited the draft. All authors have read and agreed to the published version of the manuscript.

Funding: This research was funded by NSAF (No. U1930207 and No. U1730125) and NSFC (No. 51901214).

Conflicts of Interest: The authors declare no conflict of interest.

References

1. Chen, X.; Liu, T.; Li, R.; Liu, J.S.; Zhao, Y. Molecular dynamics simulation on the shape memory effect and superelasticity in NiTi shape memory alloy. *Comput. Mater. Sci.* **2018**, *146*, 61–69. [CrossRef]
2. Luo, J.; Bobanga, J.O.; Lewandowski, J. Microstructural heterogeneity and texture of as-received, vacuum arc-cast, extruded, and re-extruded NiTi shape memory alloy. *J. Alloy. Compd.* **2017**, *712*, 494–509. [CrossRef]
3. Wang, X.B.; Verlinden, B.; Kustov, S. Multi-stage martensitic transformation in Ni-rich NiTi shape memory alloys. *Funct. Mater. Lett.* **2017**, *10*, 1740004. [CrossRef]
4. Li, J.; Wang, H.F.; Liu, J.; Ruan, J.M. Effects of Nb addition on microstructure and mechanical properties of TiNiNb alloys fabricated by elemental powder sintering. *Mater. Sci. Eng. A* **2014**, *609*, 235–240. [CrossRef]
5. Tabesh, M.; Boyd, J.; Atli, K.; Karaman, I.; Lagoudas, D. Design, fabrication, and testing of a multiple-actuation shape memory alloy pipe coupler. *J. Intell. Mater. Syst. Struct.* **2017**, *29*, 1165. [CrossRef]
6. Niccoli, F.; Garion, C.; Maletta, C.; Sgambitterra, E.; Furgiuele, F.; Chiggiato, P. Beam-pipe coupling in particle accelerators by shape memory alloy rings. *Mater. Des.* **2017**, *114*, 603–611. [CrossRef]
7. Niccoli, F.; Garion, C.; Maletta, C.; Chiggiato, P. Shape-memory alloy rings as tight couplers between ultrahigh-vacuum pipes: Design and experimental assessment. *J. Vac. Sci. Technol. A* **2017**, *35*, 031601. [CrossRef]
8. Piotrowski, B.; Ben Zineb, T.; Patoor, E.; Eberhardt, A. Modeling of niobium precipitates effect on the Ni47Ti44Nb9 Shape Memory Alloy behavior. *Int. J. Plast.* **2012**, *36*, 130–147. [CrossRef]
9. Zhu, R.; Tang, G.; Shi, S.-Q.; Fu, M. Effect of electroplastic rolling on deformability and oxidation of NiTiNb shape memory alloy. *J. Mater. Process. Technol.* **2013**, *213*, 30–35. [CrossRef]
10. Jiang, S.Y.; Mao, Z.N.; Zhang, Y.Q.; Hu, L. Mechanisms of nanocrystallization and amorphization of NiTiNb shape memory alloy subjected to severe plastic deformation. *Procedia Eng.* **2017**, *207*, 1493–1498. [CrossRef]
11. Fan, Q.; Zhang, Y.; Zhang, Y.; Wang, Y.; Yan, E.; Huang, S.; Wen, Y. Influence of Ni/Ti ratio and Nb addition on martensite transformation behavior of NiTiNb alloys. *J. Alloy. Compd.* **2019**, *790*, 1167–1176. [CrossRef]
12. Fan, Q.C.; Sun, M.Y.; Zhang, Y.H.; Wang, Y.Y.; Zhang, Y.; Peng, H.B.; Sun, K.H.; Fan, X.H.; Huang, S.K.; Wen, Y.H. Influence of precipitation on phase transformation and mechanical properties of Ni-rich NiTiNb alloys. *Mater. Charact.* **2019**, *154*, 148–160. [CrossRef]
13. Liu, W.; Zhao, X.Q. Mechanical Properties and Transformation Behavior of NiTiNb Shape Memory Alloys. *Chin. J. Aeronaut.* **2009**, *22*, 540–543.
14. He, X.M.; Rong, L.J. Effect of Deformation on the Stress-Induced Martensitic Transformation in (Ni47Ti44)100-xNbx Shape Memory Alloys with Wide Hysteresis. *Met. Mater Int.* **2006**, *12*, 279–288. [CrossRef]
15. He, X.M.; Rong, L.J. DSC analysis of reverse martensitic transformation in deformed Ti–Ni–Nb shape memory alloy. *Scr. Mater.* **2004**, *51*, 7–11. [CrossRef]
16. Wang, L.; Rong, L.J.; Yan, D.S.; Jiang, Z.M.; Li, Y.Y. DSC study of the reverse martensitic transformation behavior in a shape memory alloy pipe-joint. *Intermetallics* **2005**, *13*, 403–407. [CrossRef]
17. Tabesh, M.; Atli, K.; Rohmer, J.; Franco, B.; Karaman, I.; Boyd, J.G.; Lagoudas, D. Design of shape memory alloy pipe couplers: Modeling and experiments. *Ind. Commer. Appl. Smart Struct. Technol.* **2012**, *8343*, 83430J. [CrossRef]
18. Yang, G.L.; Ni, Z.M.; Han, J.; Wang, E.M. Simulation on Pipe Joints Expansion Technology of NiTiNb Shape Memory Alloy. *Adv. Mater. Res.* **2011**, *189–193*, 1711–1717. [CrossRef]
19. Uchida, K.; Shigenaka, N.; Sakuma, T.; Sutou, Y.; Yamauchi, K. Effects of Pre-Strain and Heat Treatment Temperature on Phase Transformation Temperature and Shape Recovery Stress of Ti-Ni-Nb Shape Memory Alloys for Pipe Joint Applications. *Mater. Trans.* **2008**, *49*, 1650–1655. [CrossRef]
20. Zhang, C.S. Effects of deformation on the transformation hysteresis and shape memory effect in a Ni47Ti44Nb9 alloy. *Scr. Mater.* **1990**, *24*, 1807–1812. [CrossRef]
21. Wang, K.L.; Lu, S.Q.; Li, G.F.; Liu, J.W.; Wang, E.M. Influence of Pre-Deformation Strain on Recovery Performance of Ni47Ti44Nb9 Alloy Φ8mm Pipe Joint. *Adv. Mater. Res.* **2015**, *1095*, 140–144. [CrossRef]
22. Li, G.F.; Lu, S.Q.; Wang, K.L.; Liu, J.W.; Wang, E.M. Influence of Predeformation Temperature on Recovery Performance of Ni47Ti44Nb9 Alloy Φ16mm Pipe Joint. *Adv. Mater. Res.* **2014**, *904*, 41–45. [CrossRef]

23. Wang, E.M.; Hong, Q.H.; Ni, Z.M.; Han, J. Influence of Processing State on Recovery Stress in NiTiNb Shape Memory Alloy. *Adv. Mater. Res.* **2014**, *875–877*, 1525–1528. [CrossRef]
24. Cai, S.; Schaffer, J.E.; Ren, Y.; Wang, L. Deformation of a super-elastic NiTiNb alloy with controllable stress hysteresis. *Appl. Phys. Lett.* **2016**, *108*, 261901. [CrossRef]
25. Hamilton, R.F.; Lanba, A.; Ozbulut, O.E.; Tittmann, B.R. Shape Memory Effect in Cast Versus Deformation-Processed NiTiNb Alloys. *Shape Mem. Superelast.* **2015**, *1*, 117–123. [CrossRef]
26. Yan, Y.; Jin, W.; Li, X. Texture Development in the Ni47Ti44Nb9 Shape Memory Alloy During Successive Thermomechanical Processing and Its Effect on Shape Memory and Mechanical Properties. *Met. Mater. Trans. A* **2012**, *44*, 978–989. [CrossRef]
27. Yin, X.; Mi, X.; Li, Y.; Gao, B. Microstructure and Properties of Deformation Processed Polycrystalline Ni47Ti44Nb9 Shape Memory Alloy. *J. Mater. Eng. Perform.* **2012**, *21*, 2684–2690. [CrossRef]
28. Feng, Z.W.; Mi, X.J.; Wang, J.B.; Yuan, Z.S.; Zhou, J. Effect of Annealing Temperature on the Transformation Temperature and Texture of Ni47Ti44Nb9 Cold-Rolled Plate. *Adv. Mater. Res.* **2012**, *557*, 1281–1287. [CrossRef]
29. Liu, H.; Sun, G.; Wang, Y.; Chen, B.; Tian, Q.; Wang, X.; Zhang, C. Texture evolution in shocked Ni47Ti44Nb9 shape memory alloys. *Mater. Sci. Technol.* **2013**, *29*, 1499–1502. [CrossRef]
30. Gall, K.; Sehitoglu, H.; Chumlyakov, Y.I.; Kireeva, I.V. Tension-compression asymmetry of the stress-strain response in aged single crystal and polycrystalline. *NiTi Acta Mater.* **1999**, *47*, 1203–1217. [CrossRef]
31. Miyazaki, S.; Kimura, S.; Otsuka, K.; Suzuki, Y. The habit plane and transformation strains associated with the martensitic transformation in Ti-Ni single crystals. *Scr. Metall.* **1984**, *18*, 883. [CrossRef]
32. Sun, G.A.; Wang, X.L.; Wang, Y.D.; Woo, W.C.; Wang, H.; Liu, X.P.; Chen, B.; Fu, Y.Q.; Sheng, L.S.; Ren, Y. In-situ high-energy synchrotron X-ray diffraction study of micromechanical behavior of multiple phases in Ni47Ti44Nb9 shape memory alloy. *Mater. Sci. Eng. A* **2013**, *560*, 458–465. [CrossRef]
33. Chen, X.; Peng, X.; Chen, B.; Han, J.; Zeng, Z.; Hu, N. Experimental investigation on transformation, reorientation and plasticity of Ni47Ti44Nb9SMA under biaxial thermal–mechanical loading. *Smart Mater. Struct.* **2015**, *24*, 75025. [CrossRef]
34. Lin, H.C.; Wu, S.K.; Chou, T.S.; Kao, H.P. The effects of cold rolling on the martensitic transformation of an equiatomic TiNi alloy. *Acta Metall. Mater.* **1991**, *39*, 2069–2080. [CrossRef]

metals

MDPI

Article

Hysteretic Behavior and Ultimate Energy Dissipation Capacity of Large Diameter Bars Made of Shape Memory Alloys under Seismic Loadings

Guillermo González-Sanz, David Galé-Lamuela, David Escolano-Margarit and Amadeo Benavent-Climent *

Department of Mechanical Engineering, Universidad Politécnica de Madrid, 28006 Madrid, Spain; guillermo.gonzalez.sanz@upm.es (G.G.-S.); david.gale@upm.es (D.G.-L.); d.escolano@upm.es (D.E.-M.)

* Correspondence: amadeo.benavent@upm.es; Tel.: +34-910-677-237

Received: 24 September 2019; Accepted: 12 October 2019; Published: 13 October 2019

Abstract: Shape memory alloys in the form of bars are increasingly used to control structures under seismic loadings. This study investigates the hysteretic behavior and the ultimate energy dissipation capacity of large-diameter NiTi bars subjected to low- and high-cycle fatigue. Several specimens are subjected to quasi-static and to dynamic cyclic loading at different frequencies. The influence of the rate of loading on the shape of the hysteresis loops is analysed in terms of the amount of dissipated energy, equivalent viscous damping, variations of the loading/unloading stresses, and residual deformations. It is found that the log-log scale shows a linear relationship between the number of cycles to failure and the normalized amount of energy dissipated in one cycle, both for low- and for high-cycle fatigue. Based on the experimental results, a numerical model is proposed that consists of two springs with different restoring force characteristics (flag-shape and elastic-perfectly plastic) connected in series. The model can be used to characterize the hysteretic behavior of NiTi bars used as energy dissipation devices in advanced earthquake resistant structures. The model is validated with shake table tests conducted on a reinforced concrete structure equipped with 12.7 mm diameter NiTi bars as energy dissipation devices.

Keywords: shape memory alloys; cyclic tests; fatigue test; energy dissipation; earthquake engineering

---

1. Introduction

Earthquakes cause heavy casualties and property damage worth billions. Designing structures to withstand the vibrations induced by seismic actions is of primary concern. The conventional seismic design approach relies on the ability of a structure to dissipate the energy input by the earthquake through inelastic deformations in special regions of the structure (e.g., column bases and beam-ends in frame structures). This conventional approach accepts heavy damage (or the need for possible demolition) on the structure after a severe earthquake. To overcome this drawback and improve overall performance, an important research effort in the last two decades has been devoted to the development of innovative structures with passive control systems. This type of structure features special members called energy dissipation devices that are designed to dissipate most of the energy input by the earthquake, thus minimizing or avoiding damage to the main structure. Among the different types of energy dissipation devices developed in the past, those based on the use of shape memory alloys (SMAs) are particularly appealing because, in addition to dissipating energy, they are able to regain their original shape after being deformed well beyond 6–8% strain [1–8]. This ability is a result of a phase transformation that may be induced by either stress or a temperature change. SMAs have a crystal structure with two main phase transformations: the martensite and the austenite. Under high stresses detwinned martensite is the stable crystal structure. Twinned martensite may exist as a pure crystal structure under low stress levels. The austenite phase is stable at high temperatures or low stresses. The phase transformation involves four characteristic temperatures: M_f, M_s, A_s and A_f (ordered from lowest to highest). During the forward transformation, under zero load, austenite begins to transform to twinned martensite at the martensitic start temperature M_s. This transformation completes to martensite at the martensitic finish temperature M_f. At this stage, the material is fully in the twinned martensitic phase. During heating, the reverse transformation initiates at the austenitic start temperature A_s and the transformation is completed at the austenitic finish temperature A_f [9]. Figure 1 shows a typical phase diagram and stress-strain-temperature curve of a NiTi SMA [10].

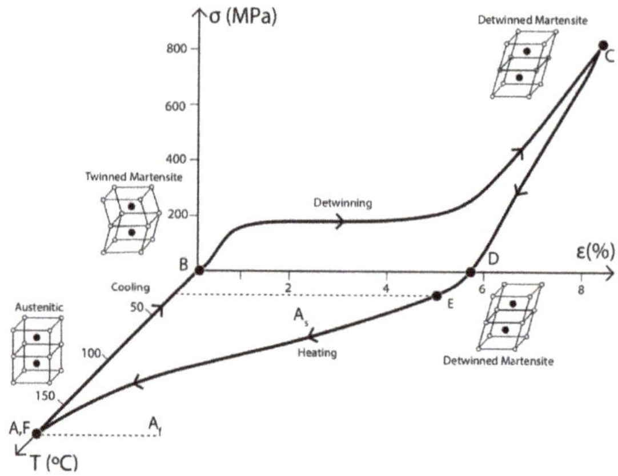

Figure 1. Phase diagram and stress-strain-temperature curve.

The key advantage of this solid-solid transformation is the reversibility of the process and the complete recovery of strain. In most civil engineering applications, SMAs are used at temperatures above A_f, with the material in its austenite phase. When sufficiently high stress is applied to the material in the austenite phase, the SMA transforms into so-called "detwinned" martensite. When the load is released, a reverse transformation to the austenite state occurs, resulting in nearly complete shape recovery and a substantial hysteretic loop. The shape recovery is known as the superelastic effect, and it provides the structures equipped with SMAs with recentering properties. The hysteretic loop is a source of energy dissipation. The mechanical behavior of superelastic SMAs fits perfectly with the requirements of a seismic control device [2]. The main benefits can be summarized as follows: (i) reduction or even nullification of the residual deformation on the main structure after the earthquake due to the self-centering property, (ii) increase in the energy dissipation capacity of the overall structure, (iii) limitation of the forces imparted to the main structure because of the stress plateau present in strain levels up to 6–8%, (iv) reduction of lateral displacements and, as a result, limitation of the *P*-Δ effects, and (v) excellent resistance to corrosion and high-cycle resistance. The P-Δ the effect is a destabilizing moment that takes place when the structure deforms laterally (e.g., due to earthquake or wind loads), which equals the force of the gravity loads multiplied by the horizontal displacement of the structure. The most commonly used SMAs are those based on nickel-titanium- (NiTi) and copper- (Cu) based alloys [2,3,7]. For engineering applications, the almost equiatomic system of NiTi alloys is found to be the best combination, owing to their temperature variation stability and higher resistance to corrosion and fatigue [1,3].

Figure 2 shows the typical stress-strain relationship of a bar made of SMA subjected to a cycle of forced displacements at constant temperature, together with several parameters that characterize its mechanical behavior [3,4]: residual strain ε_R, loading transformation stresses at start σ_{Ls} and finish σ_{Lf} of the phase, unloading transformation stresses at start σ_{ULs} and finish σ_{ULf} of the phase, initial elastic modulus E_A, loading phase transformation elastic modulus $E_{A\text{-}M}$, unloading phase transformation elastic modulus $E_{M\text{-}A}$, and modulus in the full martensite phase E_M. The shaded area in Figure 2 represents the energy dissipated in one loop of hysteresis E_D. To determine σ_{Ls}, σ_{Lf}, σ_{ULs}, and σ_{ULf}, five lines tangent to the relevant parts of the ε-σ curve are drawn, as shown with dot lines in Figure 2. The slopes of these lines are E_A, $E_{A\text{-}M}$, E_M, $E_{M\text{-}A}$, and E_A, respectively. σ_{Ls}, σ_{Lf}, σ_{ULs}, and σ_{ULf} are the ordinates of the intersection points of these lines.

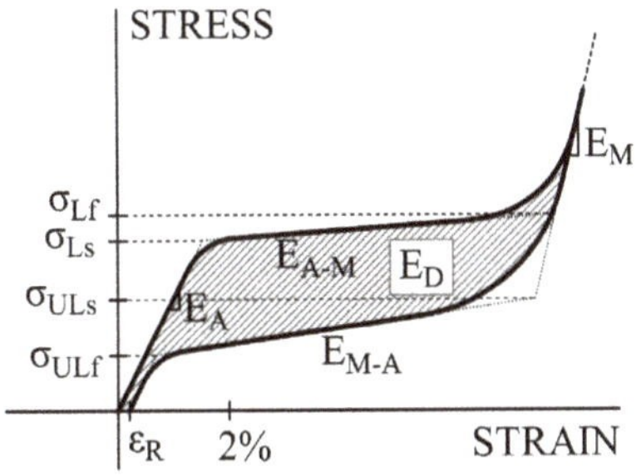

Figure 2. Idealized cyclic behavior of superelastic shape memory alloy (SMA).

Residual strain is a measure of the recentering capability, and it refers to the strain that the material does not recover when it returns to a zero stress state. The points that define the loading and unloading transformation stresses are not evident because of the nonlinear shape of the hysteresis loops. A simplified approximation is to define the loading transformation stress at the 2% strain from the initial loading cycle [3]. Nevertheless, in order to define the four transformation stress points, a better approximation is to estimate the change of slope on the curve for loading and unloading paths. E_A is referred to as the austenite stable phase and it provides valuable information to calculate the initial stiffness of the device. Two additional parameters that characterize the mechanical behavior of bars made of NiTi alloys are the equivalent viscous damping ratio associated with a given cycle ζ_{eq} and number of cycles to failure. For a given cycle, ζ_{eq} is defined by [11]:

$$\zeta_{eq} = \frac{E_D}{2\pi\varepsilon_{max}\sigma_{max}V} \tag{1}$$

where ε_{max} and σ_{max} are the maximum strain and stress, respectively, in the cycle, and V is the volume of the bar. The number of cycles to failure is an indicator of the life of the material.

Past research showed that hysteretic behavior and the superelastic properties of SMAs vary during the cyclic loading. For cycles of constant amplitude and during the first cycles there is a decrease in σ_{Ls} and σ_{Lf}, while σ_{ULs} and σ_{Ulf} remain approximately constant. This results in a decrease of the energy E_D dissipated in the cycle and in the corresponding ζ_{eq}. However, the response stabilizes as the number of cycles increases. Under cycles of increasing amplitude, ε_R remains constant as well as σ_{Ls} and σ_{Lf}, whereas σ_{ULs} and σ_{Ulf} decrease. Consequently, the hysteresis shape involves greater energy dissipated, resulting in higher equivalent viscous damping. When the loading stress plateau (i.e., the segment with slope $E_{A\text{-}M}$ in Figure 2) is overcome (onset of the pure martensite phase), the strain-stress curve exhibits a strain hardening effect [1–5].

The hysteretic behavior and superelastic properties of SMAs are influenced by the size, strain rate, and temperature. As for the size effect, the superelastic properties can be achieved both in bars and in wires, although some studies reveal less residual strain in large diameter bars [4]. Loading transformation stresses are lower in bars than in wires and unloading transformation stress becomes higher in bars. The consequence is a narrower hysteresis shape, with less energy dissipated per cycle for bars in comparison with wires, and a lower equivalent viscous damping. It has been shown that the strain rate effect does not have a remarkable impact on the residual strain ε_R but does indeed have a significant influence on the shape of the hysteresis loops. Increasing the strain rate results in a vertical displacement (i.e., both loading and unloading transformation stresses increase) and narrowing of the hysteretic loops. The narrowing implies a loss of energy dissipated per cycle. This reduction of dissipated energy, together with the vertical displacement of the loops, reduces the equivalent viscous damping. The reason for this behavior is the self-heating of the material associated with an increasing difficulty to transfer the heat generated between phase transformations at high strain rates [1–3].

Most past studies have focused on small diameter wires, and there is little information on large diameter bars, which are more usual in earthquake engineering applications. Past research did not clarify if the strain rate effect varies significantly with size. As for the effect of temperature, it has been shown that the residual strain remains the same for temperatures larger than A_f, thus keeping the recentering capability unaffected. However, increasing temperature above A_f causes a vertical displacement of the hysteresis loop (i.e., higher loading and unloading transformation stresses), while the energy dissipated per cycle remains almost the same. The vertical displacement of the loops, despite the fact that E_D remains unchanged, results in a reduction of ζ_{eq}. Finally, regarding the amount of energy that the SMA can dissipate up to failure (ultimate energy dissipation capacity), most studies to date address this issue as a problem of high-cycle fatigue [12]. Yet earthquakes impose on the structures a relatively low number of cycles (in comparison with wind or traffic loads) having high stress levels that involve plastic deformations. These are the two common factors attributed to low-cycle fatigue. Some recent investigations on SMA low-cycle fatigue have been carried out on small diameter wires and micro-tubes [13,14], whose conclusions may not be consistent with large diameter bars. Studies on the ultimate energy dissipation capacity of SMA large diameter bars under low cyclic fatigue are almost inexistent.

This paper presents an experimental study aimed at (i) characterizing the hysteretic behavior and evaluating the ultimate energy dissipation capacity of large diameter bars made of NiTi alloys, and (ii) proposing a simple numerical model that can be easily implemented in a finite element code to perform non-linear time history analyses and obtain the seismic response of structures equipped with devices that use large diameter NiTi bars as a source of energy dissipation. First, the results of quasi-static and dynamic cyclic tests conducted on NiTi bars isolated from the structure are presented. Next, a simple numerical model for characterizing their hysteretic behavior is developed and calibrated with the results of these tests. Finally, the NiTi bars were assembled to form an energy dissipation device that was installed in a reinforced concrete (RC) structure and subjected to seismic simulations in a shake table. The results of the shake table tests were used to validate the numerical model proposed.

2. Cyclic Tests on NiTi Bars

2.1. Test Specimens

Initially, thirteen specimens consisting of cylindrical bars, 12.7 mm in diameter and 750 mm in length, made of SMAs were tested under quasi-static and dynamic cyclic loadings. All the specimens have the same geometry, material composition, and thermo-mechanical processing. The bar was heat treated for superelastic properties. The heat treatment applied guaranteed the superelastic properties at room temperature by full annealment with a reference temperature (A_s) between −30 and −10 °C. The transition temperatures are M_f = −37.66 °C, M_s = −31.36 °C, A_s = −16.13 °C and A_f = −5.16 °C. Figure 3 shows the results of the (Differential Scanning Calorimetry) DSC experiment. The NiTi bars were manufactured by the company SAES Smart Materials (New Hartford, NY, USA). Unfortunately, no more specific material-related information was available from the manufacturer.

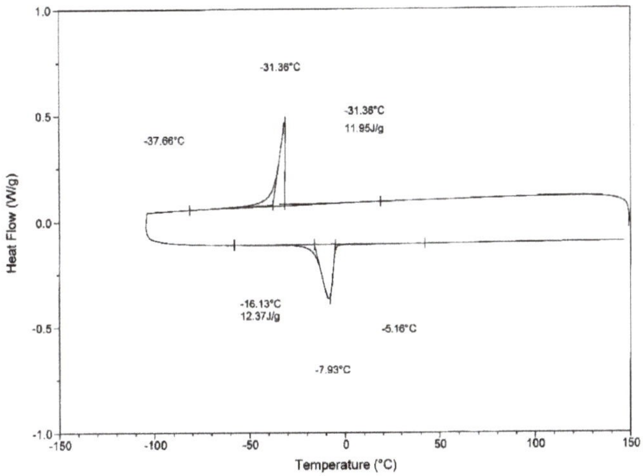

Figure 3. DSC experiment.

2.2. Loading Set up and Loading Protocol

The authors considered the standard ASTM F20516 ("Standard Test Method for Tension Testing of Nickel-Titanium Superelastic Materials") and testing protocols used in the literature [1–5]. Herein, the tests conducted at frequencies less than 0.05 Hz are considered as quasi-static and those conducted at frequencies larger than 0.2 Hz as dynamic.

2.2.1. Quasi-Static Cyclic Tests

A total of eleven specimens were tested under quasi-static loads, applying different frequencies and loading patterns, as explained below. Six specimens were subjected to cyclic displacements of constant amplitude at $\varepsilon = 4.5\%$ with a frequency of 0.02 Hz, following the loading protocol 1 shown in Figure 4. These specimens will be referred to as S_{11} to S_{16}, where the letter refers to Static, the first sub index identifies the loading protocol, and the second sub index the number of the specimen. The results of the SMA under constant amplitude and low frequency test (protocol 1) were used as "benchmark response". A relatively large number of specimens were tested with this protocol 1 in order to assess the repeatability of the results, particularity in terms of (i) number of cycles required to stabilize the shape of the hysteresis loops, (ii) the maximum stress attained in the cycle, and (iii) the residual strain. The coefficients of variation (ratio of standard deviation to the mean) obtained for these variables were 0, 0.04 and 0.10, respectively. Two specimens (referred to as S_{21} to S_{22}) were subjected to cyclic loads following the loading protocol 2 shown in Figure 5, applied at two different frequencies of 0.02 Hz (in specimen S_{21}) and 0.04 Hz (in specimen S_{22}). Two specimens (referred to as S_{31} and S_{32}) were subjected to cyclic displacements until failure, following the multiple-step loading protocol 3 shown in Figure 6 at a frequency of 0.02 Hz (quasi-static). One specimen (referred to as S_{41}) was subjected to the cyclic displacements until failure following the loading pattern shown in Figure 7 applied at a frequency of 0.02 Hz (quasi-static). All tests were conducted in ambient conditions (20–25 °C) with a universal testing machine, SAXEWAY T1000 (MOOG Inc., East Aurora, NY, USA). The experimental set up involved a pair of transducers for the displacement control plus the internal load cell of the actuator that measured the applied force.

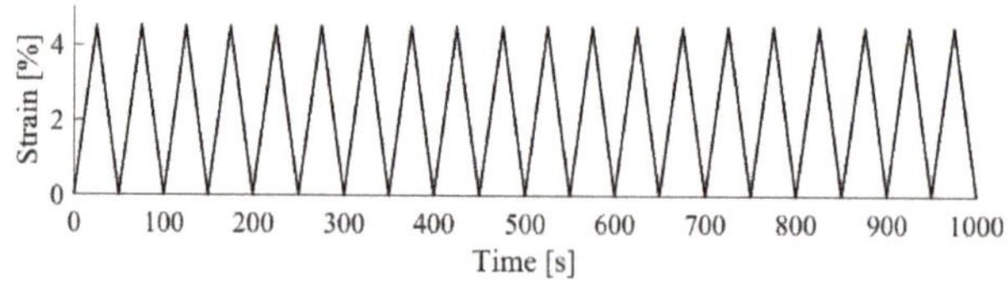

Figure 4. Loading protocol 1 (frequency 0.02 Hz) used for specimens S_{11} to S_{16}.

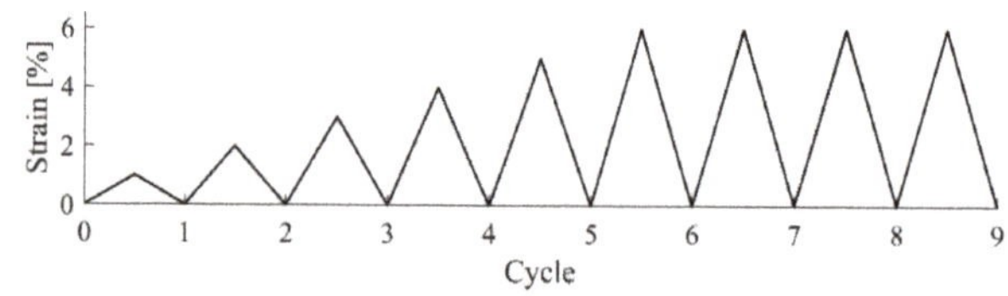

Figure 5. Loading protocol 2 applied at frequencies 0.02, 0.04, 0.2, 1.0 Hz used for S_{21}, S_{22}, D_{21}, D_{22}.

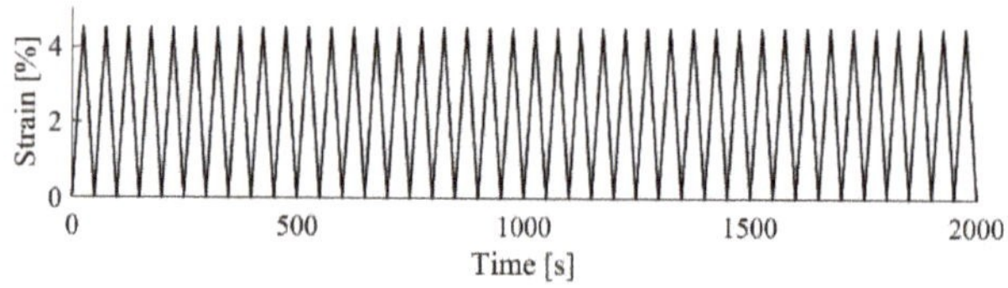

Figure 6. Loading protocol 3 (frequency 0.02 Hz) used for specimens S_{31}, S_{32}.

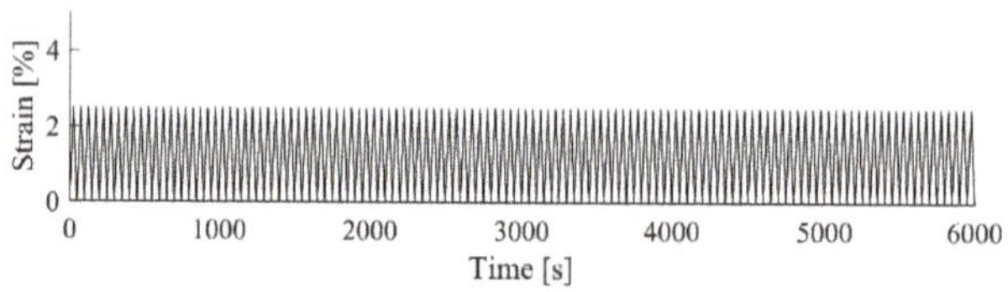

Figure 7. Loading protocol 4 (frequency 0.02 Hz) used for specimen S_{41}.

2.2.2. Dynamic Cyclic Tests

Two additional specimens were tested under dynamic loads, following the loading protocol shown in Figure 5 applied at two different frequencies, 0.2 Hz and 1.0 Hz. These specimens will be referred to hereafter as D_{21} and D_{22}, respectively, where the letter indicates Dynamic. The tests were conducted in ambient conditions (20–25 °C) with an INSTRON 8803 fatigue testing system (INSTRON, Norwood, MA, USA).

2.3. *Test Results and Discussion*

2.3.1. Hysteretic Behavior

Figure 8 shows the stress-strain curves obtained for specimen S_{11}. Specimens S_{12} to S_{16} exhibited similar behavior. Three relevant features should be noted. First, in the initial cycles, the loading and unloading transformation stresses tend to diminish, which results in a reduction of the energy dissipated in each cycle. This phenomenon is called in the literature "functional fatigue" [15]. In successive cycles, the shape of the hysteresis loops tends to quickly stabilize, becoming almost identical. Second, the maximum stress is practically the same in all cycles. Third, the residual strain ε_R remains approximately constant.

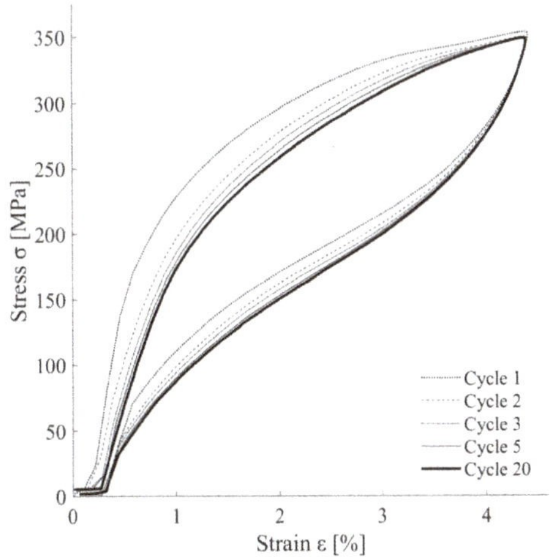

Figure 8. Hysteresis loops of specimen S_{11}.

Figure 9 shows the stress-strain curves obtained for specimens S_{21}, S_{22}, D_{21}, and D_{22}, subjected to cycles of increasing amplitude following protocol 2 at frequencies of 0.02, 0.04 (quasi-static) and 0.2, 1.0 Hz (dynamic), respectively. The first frequency (0.2 Hz) is approximately the fundamental frequency of a high-rise building vibrating in the fundamental mode. The second frequency (1 Hz) is typical of low to moderate rise buildings. Two relevant features should be noted. First, for a fixed frequency, the shape of the loops at different amplitudes is seen to follow basically the same pattern, that is, the loading and unloading paths for a given amplitude overlap the loading and unloading paths obtained in cycles of lower amplitude. Kimiecik et al. [16] studied the configurations of transforming martensite during ambient temperature cyclic deformations of superelastic NiTi and found that local transformation history is responsible for this macroscopically observed performance.

Second, the shape of the loops differs notably depending on the frequency of the loading. In order to better understand the effect of the strain rate, the hysteresis loops at 6% strain described by each specimen are shown in Figure 9. It is clear from this Figure that an increase in the strain rate results in (i) greater values for loading and unloading transformation stresses, (ii) narrower hysteresis loops, and (iii) earlier and more remarkable occurrence of the strain hardening effect. These variations imply a decrease in the energy dissipation per cycle and in the equivalent viscous damping as the frequency of the applied loads increases. The obtained response is justified by the high dependency of SMA response on thermo-mechanical loading conditions [17–19]. More precisely, the self-heating of the SMA under cyclic loading and the difficulty of transferring this heat to the environment at high strain rates is responsible for the alterations observed on the shape of the hysteretic loops. It can be also observed in Figure 10 that the amount of residual strain is not clearly affected by the frequency of loading.

Figure 11 shows the equivalent viscous damping ratios computed for specimens S21, S22, D21, and D22. It is worth recalling that specimens S21, S22, and D21 were subjected to one cycle at amplitudes $\varepsilon = 1\%, 2\%, 3\%, 4\%$, and 5%, and to four cycles at amplitude $\varepsilon = 6\%$. Therefore, each specimen provided four values of ζ_{eq} at $\varepsilon = 6\%$. Yet specimen D22 failed during the second cycle at $\varepsilon = 6\%$, therefore there is only one point at this strain amplitude. It can be observed that, for the same frequency, ζ_{eq} tends to increase with ε, and becomes approximately constant beyond $\varepsilon = 4\%$. Under cycles of constant amplitude at $\varepsilon = 6\%$, ζ_{eq} tends to decrease with the number of cycles applied.

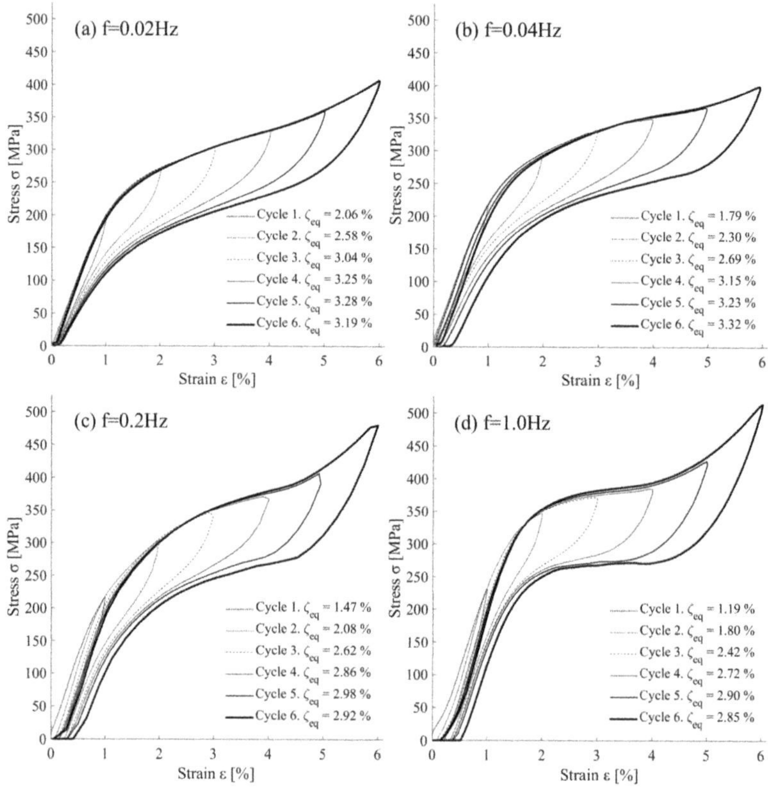

Figure 9. Hysteresis loops of specimens: (**a**) S_{21}, (**b**) S_{22}, (**c**) D_{21}, and (**d**) D_{22}.

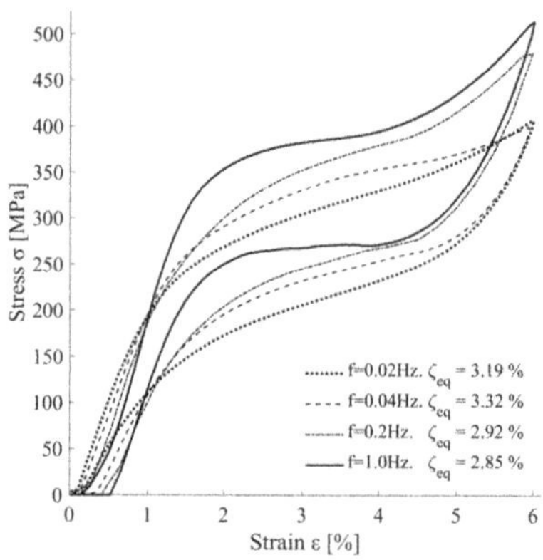

Figure 10. Hysteresis loops of specimens under quasi-static and dynamic loading at 6% strain.

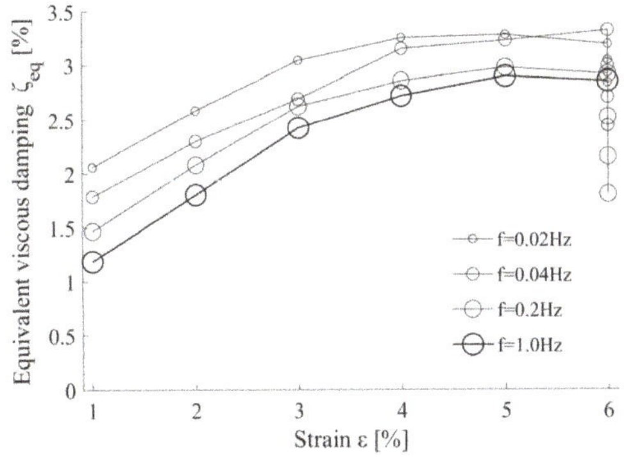

Figure 11. Equivalent damping ζ_{eq} for 12.7 mm diameter NiTi bars.

The effect of the strain rate on the amount of energy dissipated in a single cycle at $\varepsilon = 6\%$ amplitude was further investigated to identify possible differences between wires and bars made of NiTi alloys. To this end, the results obtained in this study with specimens S_{21} and D_{22} were compared with the tests conducted by other researchers [2–4,17] on wires and bars subjected to one cycle of amplitude $\varepsilon = 6\%$ at frequencies of 0.02 Hz and 1.0 Hz. The information regarding the SMA materials used in these studies can be summarized as follows. In [2]: NiTi 50%Ni, cold working and annealing. In [3] NiTi 56%Ni, cold drawn with 30% cold working and annealing. In [4]: NiTi near equiatomic, cold drawn 30% and cold worked prior to annealing. In [17]: Nitinol.

In order to make the results comparable, the energy E_D dissipated in one cycle was normalized by the product of the yield force F_y and yield displacement δ_y determined as follows. The loading branch of the force-deformation curve F-δ obtained experimentally was idealized with two segments as shown in Figure 12. The slope and position of these segments were determined so that: (i) the slope of the second segment of the bilinear approximation closely fits the path of the loading transformation phase, and (ii) the area under the real curve and the bilinear approximation was the same. The results are shown in Table 1. In this Table, ϕ is the diameter of the wire or bar, σ_y is the yield stress obtained dividing F_y by $\pi(\phi/2)^2$, ε_y is the yield strain obtained dividing δ_y by the initial length, and $\overline{E}_D$ is the dissipated energy E_D normalized by $Fy\delta y$. The values of σ_y, ε_y, E_A, and $\overline{E}_D$ vary depending on the frequency applied. The last column of Table 1 shows the ratio between $\overline{E}_D$ obtained for quasi-static loading (0.02Hz), $\overline{E}_{D,static}$, and the corresponding value obtained for dynamic loading (1.0 Hz), $\overline{E}_{D,dynamic}$. The specimens with $\phi < 2$ mm are referred to as wires hereafter, and those with $\phi > 6$ mm as bars.

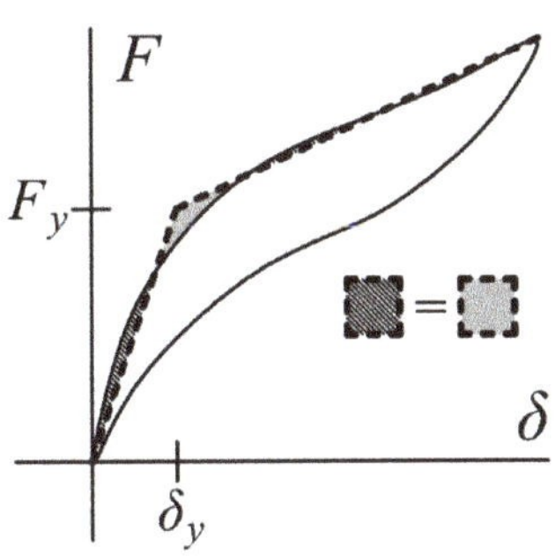

Figure 12. Idealization of the loading branch of the force-deformation curve with two segments.

Table 1. Energy dissipated in a single cycle at ε = 6% amplitude and different frequencies.

Reference	Frequency (Hz): ϕ (mm)	σ_y (MPa)		ε_y (%)		E_A (MPa)		$\overline{E}_D$=$E_D/(F_y\delta_y)$		$\frac{\overline{E}_{D,dynamic}}{\overline{E}_{D,static}}$
		0.02	1.0	0.02	1.0	0.02	1.0	0.02	1.0	
McCormick [4]	0.25	504	553	1.83	1.79	275	309	1.12	1.00	0.89
Zhu [17]	0.58	305	309	1.22	1.12	250	275	2.10	1.48	0.71
Dolce [2]	1.84	390	415	1.40	1.25	279	332	1.00	0.83	0.84
DesRoches [3]	7.10	315	374	1.33	1.40	237	267	1.67	0.71	0.42
McCormick [4]	12.70	328	414	1.33	1.54	247	269	1.41	0.78	0.56
This study	12.70	245	359	1.12	1.39	219	258	1.78	1.11	0.62

For the same type of loading (i.e., quasi-static or dynamic), Table 1 does not reflect any clear differences between wires and bars in terms of normalized dissipated energy, the average normalized energy under quasi-static loads, $\overline{E}_{D,static}$, is 1.4 in wires and 1.62 in bars, while the counterpart values under dynamic loadings, $\overline{E}_{D,dynamic}$, are respectively 1.1 and 0.87. Still, there is a clear difference in terms of the reduction of normalized energy dissipation due to the type of loading. More precisely, $\overline{E}_{D,dynamic}/\overline{E}_{D,static}$ is about 40% greater on bars than on wires. It is also seen that the yielding stress σ_y increases with the frequency. This increase is larger in bars than in wires. As for the yielding strain, it is reduced in wires when the load is applied dynamically in comparison with the static case. The tendency is the opposite for bars, however. This is possibly due to the sample size that may produce different deformation mechanisms within the superelastic strain range [20].

2.3.2. Ultimate Energy Dissipation Capacity

Past research on the fatigue of NiTi SMAs distinguishes their functional fatigue from structural fatigue [15,21]. Functional fatigue was described in Section 2.3.1. Structural fatigue refers to the gradual loss of strength under repeated loading that occurs after applying a large number of cycles, driving the specimen to failure. In turn, two types of structural fatigue can be distinguished: high-cycle fatigue and low-cycle fatigue. The former occurs after a relatively large number of cycles (several thousands of cycles) in which the material remains in the elastic range. The latter occurs at a small number of cycles (several dozens or hundreds) and involves plastic deformations. The vibrational response of a structure subjected to severe earthquakes is characterized by an occurrence from several dozens to several hundreds of cycles in which the members are strained beyond the elastic range. Therefore, the problem of fatigue in structural members subjected to seismic loadings is a problem of low-cycle fatigue.

Some recent works aspire to gain a better understanding of the structural fatigue of NiTi SMAs based on both experimental and theoretical investigations [14,22,23]. Several fatigue failure models have been developed following an energy criterion that appears to keep a good correlation between experimental and prediction approaches. However, failure models are based on tests conducted with NiTi wires and micro-tubes. This section investigates their validity for NiTi bars. To this end, the results of the tests conducted on specimens S_{31}, S_{32}, and S_{41} subjected to cycles of constant amplitude were used. The total energy dissipated by these specimens and accumulated in successive cycles until failure, $\sum \overline{E}_{D,static}$, was normalized by $F_y\delta_y$ (determined as described in Section 2.3.1), i.e., $\sum \overline{E}_{D,static} = \sum E_{D,static}/(F_y\delta_y)$, and is shown in the second column of Table 2. The normalized energy dissipated in the first cycle was also calculated for each specimen, i.e., $\overline{E}_{D,static} = E_{D,static}/(F_y\delta_y)$, and is shown in the third column of Table 2. Further, the total amount of dissipated energy was expressed in terms of the equivalent number of cycles N_f defined by $N_f = \sum \overline{E}_{D,static}/\overline{E}_{D,static}$ and is shown in the fourth column of Table 2. It is worth recalling that the loads applied to specimens S_{31}, S_{32} and S_{41} were quasi-static, and it was shown in Section 2.3.1 that the amount of energy dissipated under dynamic loading is smaller than under static loads. More precisely, for the 12.7 mm diameter bars tested in this study, the ratio $\overline{E}_{D,dynamic}/\overline{E}_{D,static}$ is 0.62 (last row in Table 1). Therefore, the normalized energy dissipated in a single cycle under dynamic loading $\overline{E}_{D,dynamic}$ can be estimated by multiplying $\overline{E}_{D,static}$ by 0.62, and it is indicated in the last column of Table 2. The pairs of values (N_f, $\overline{E}_{D,dynamic}$) obtained in this way for specimens S_{31}, S_{32} and S_{41} are plotted with circles in Figure 13 and compared with those obtained by [24] (square symbols) for 6.2 mm diameter NiTi bars tested under dynamic (0.3 Hz) loads. Since the ratio $\overline{E}_{D,dynamic}$ increases with the amplitude of the cycle, $\overline{E}_{D,dynamic}$ is directly related to the amplitude of the cyclic loading. The ultimate energy dissipation capacity corresponding to each point, shown in Figure 13, can be simply obtained by multiplying its abscissa (N_f) by its ordinate $\overline{E}_{D,dynamic}$.

Table 2. Ultimate normalized energy dissipation capacity.

ε (%)	$\sum\overline{E}_{D,static}$	$\overline{E}_{D,static}$	N_f	$\overline{E}_{D,dynamic}$
4.0	124	1.17	106	0.73
4.0	112	1.17	96	0.73
2.5	260	0.52	500	0.32

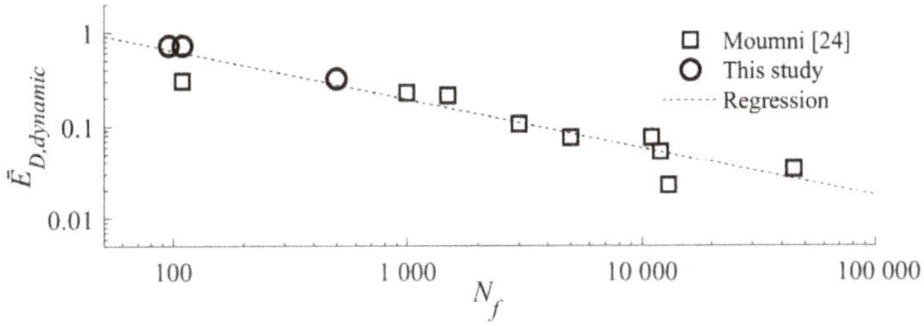

Figure 13. Normalized energy dissipation versus the number of cycles to failure.

Figure 13 shows that the ultimate energy dissipation capacity of NiTi bars under low-cycle fatigue obtained in this study is consistent with the values reported in [24]. Performing a regression analysis of these results, the ultimate energy dissipation capacity of NiTi bars subjected to dynamic cyclic loading, expressed in terms of N_f and $\overline{E}_{D,dynamic}$, can be approximated through the following expression:

$$\overline{E}_{D,dynamic} = 7.0N_f^{-0.52} \tag{2}$$

3. Numerical Characterization of the Hysteretic Behavior of NiTi Bars

Several models have been proposed in the literature to characterize the hysteretic behavior of NiTi bars. Ikeda et al. [25] proposed a specimen-based macroscopic model of SMA for unidirectional loading that considers the inner hysteretic loops of a stress-strain-temperature relationship and includes the memory effect of deformation history. This model was derived from a grain-based microscopic model. Saleeb et al. [26] proposed a fully general, three-dimensional, constitutive model for SMA that can describe all the salient features of SMA evolutionary response under complex thermo-mechanical loading conditions. This model uses multiple inelastic mechanisms to organize the exchange between the energy stored and energy dissipated during the deformation history. Karakalas et al. [27] proposed a different approach that combines a physical constitutive model with an expression that describes the hardening behavior of SMA.

Because of its simplicity, the conventional flag-shape model displayed in Figure 14 has been widely used to represent the hysteretic behavior—i.e., axial force F versus axial deformation δ—of NiTi bars. In the conventional flag-shape model the behavior of the SMAs is defined by several parameters: initial stiffness k_1, loading phase transformation stiffness k_{2L}, unloading phase transformation stiffness k_{2UL}, the strain hardening stiffness k_3, loading transformation strength at start F_{Ls}, loading transformation deformation at finish δ_{Lf}, and ratio of loading transformation strength at start β. Here, F_{Ls} and k_1 can be easily determined from the geometry (cross area A and length L) of the bar, and the mechanical properties of the material (Young's modulus E and yield stress σ_{LS}), i.e., $F_{LS} = \sigma_{Ls}A$ and $k_1 = EA/L$. The simplicity reduces computational efforts substantially when performing complex time history nonlinear analysis of structures subjected to seismic loadings. The conventional flag-shape model cannot, however, capture the residual deformation associated with the residual strain ε_R typically exhibited by all hysteresis loops, as seen in Figure 10. If the NiTi bar is subjected to just a few cycles of large amplitude (e.g., far beyond $\varepsilon = \sigma_{LS}/E_A$ in Figure 2), the amount of dissipated energy associated with the residual strain ε_R (represented by the area $\sigma_{LS}\ \varepsilon_R$ in Figure 2) is negligible in comparison with the energy dissipated in a complete cycle (represented by the shaded area of the complete loop in Figure 2). In such a case, the conventional flag-shape model captures the actual amount of energy dissipated by the NiTi bar reasonably well. Yet if the loading history consists of a combination of few cycles of large amplitude and a large number of cycles of small amplitude, i.e., below $\varepsilon = \sigma_{LS}/E_A$ in Figure 2, the amount of energy dissipated by the small amplitude cycles can be comparatively large. The latter is the typical displacement pattern imposed by earthquakes on structural members. In this case, the conventional flag-type model can lead to a wrong prediction of the energy accumulated on the NiTi bars and to an unsafe estimation of failure.

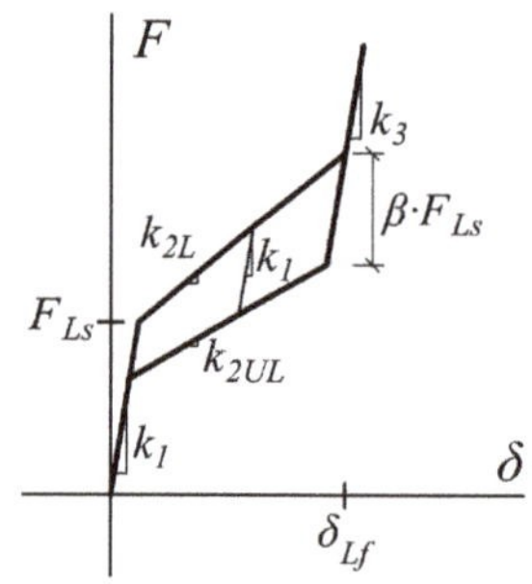

Figure 14. Conventional flag-shape model.

To remediate this flaw without sacrificing simplicity, a new hysteretic model is proposed for characterizing the hysteretic behavior of NiTi bars. It consists of two springs connected in parallel. The force-displacement relationship of one of the springs follows the conventional flag-type model described above and shown in Figure 14, but with the initial stiffness and the loading transformation strength at start weighted by a factor γ, as shown in Figure 15a. The second spring follows an elastic-perfectly plastic (EPP) hysteretic behavior, with yield force equal to $(1-\gamma)F_{LS}$ and elastic stiffness k_{EPP}, as shown in Figure 15b. The sum of the restoring forces provided by each spring at a given displacement δ gives the complete hysteretic model depicted in Figure 15c. Worth noting in Figure 15c is that the secant stiffness at $\delta = \delta_{LS}$ gives the stiffness $k_1 = EA/L$.

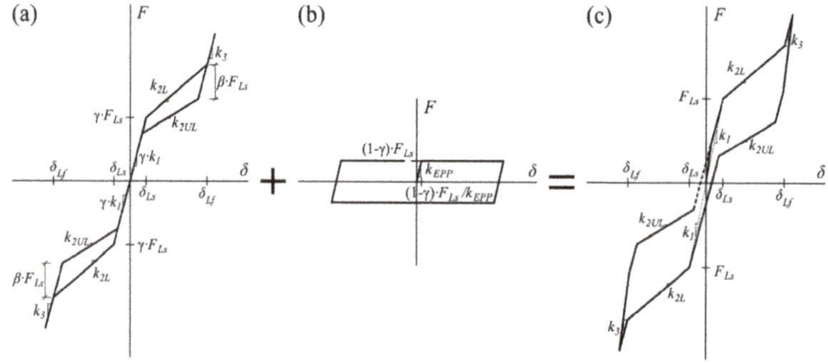

Figure 15. Proposed hysteretic model: (**a**) flag-shape component, (**b**) elastic-perfectly plastic component, (**c**) complete model.

In the proposed hysteretic model, F_{LS} and k_1 are determined from the geometry of the bar and the mechanical properties of the material, as indicated above (i.e., $F_{LS} = \sigma_{LS}A$ and $k_1 = EA/L$). The rest of the parameters were calibrated with the results of the dynamic cyclic tests described in Section 2, giving: $k_{2L} = k_1/15$, $k_{2UL} = 3k_1/50$, $k_3 = k_1$, $\delta_{Lf} = 4F_{LS}/k_1$, $\beta = 0.1$, $k_{EPP} = 4k_1$, and $\gamma = 0.86$. Figure 16 compares the shape of the hysteresis loops obtained with the proposed model and the results of the dynamic cyclic tests (specimen D_{22}). Comparison in terms of dissipated energy gives $E_D = 130$ kN·mm for the numerical model and $E_D = 119$ kN·mm for the test, the difference being less than 10%.

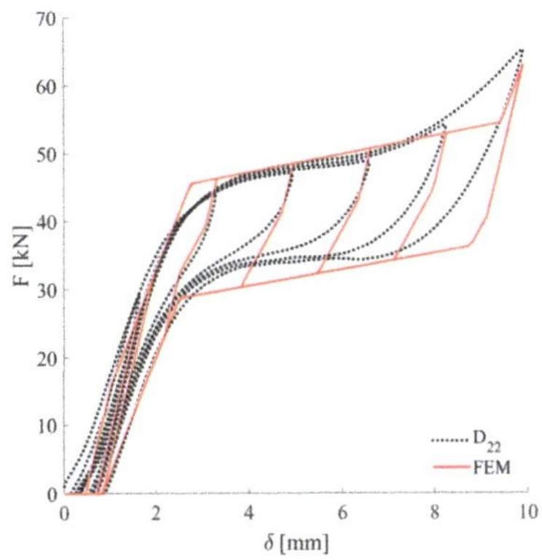

Figure 16. Comparison between the proposed hysteretic model and dynamic cyclic tests.

4. Shake Table Tests of a Structure with NiTi Bars

4.1. Brace-Type NiTi Damper

In order to validate the hysteretic model proposed in Section 3, several NiTi bars identical to those described in Section 2 were assembled forming the hysteretic damper shown in Figure 17. The damper has the form of a conventional brace and is intended to be installed in a structure as a standard diagonal bar. It is constructed by assembling two standard hollow structural rectangular sections, one into the other, with a central NiTi bar. The inner tube has a central steel plate. The NiTi bar is fixed at mid-length to the central steel plate and at both ends to the inner and outer tubes, with the mechanical anchors shown in Figure 17c. The NiTi bar is arranged inside the tubes in such a way that when the damper is subjected to forced axial deformations (tension or compression), there is always one half of the NiTi bar of length L carrying tension forces, while the other half does not carry any load.

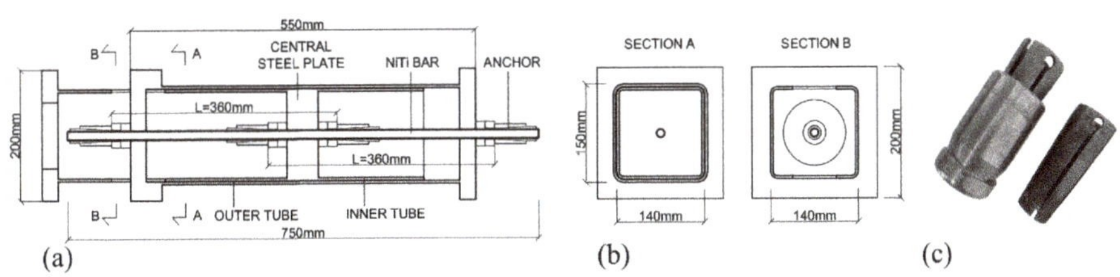

Figure 17. Assemblage of NiTi bars to form a hysteretic damper: (**a**) elevation, (**b**) sections, (**c**) detail of the anchorage of the NiTi bars.

4.2. Test Specimen and Experimental Set up

First, a prototype reinforced concrete (RC) structure—consisting of waffle flat plates supported on isolated columns—was designed. The prototype had three stories and the dimensions in the plan were 18 × 15 m^2. Second, a partial structural model having three columns and the height of one story and a half was selected from this prototype structure. Third, a test specimen was defined from the partial structural model by applying scale factors of $\lambda_L = 2/5$ for length. The test specimen was built in Laboratory and three brace-type hysteretic dampers consisting of NiTi bars and steel tubes assembled as shown in Figure 17 were installed in each story as diagonal elements. Additional steel blocks were attached at the top of the RC plate and at the top half of the columns of the second story to represent the gravity loads acting on the floors. Finally, the test specimen with the brace-type NiTi dampers was mounted on a bidirectional 3 × 3 m^2 shake table forming the experimental set up shown in Figures 18 and 19. The brace-type NiTi dampers were instrumented with displacement transducers and strain gauges.

Figure 18. Overall view of the experimental set up for the shake table tests.

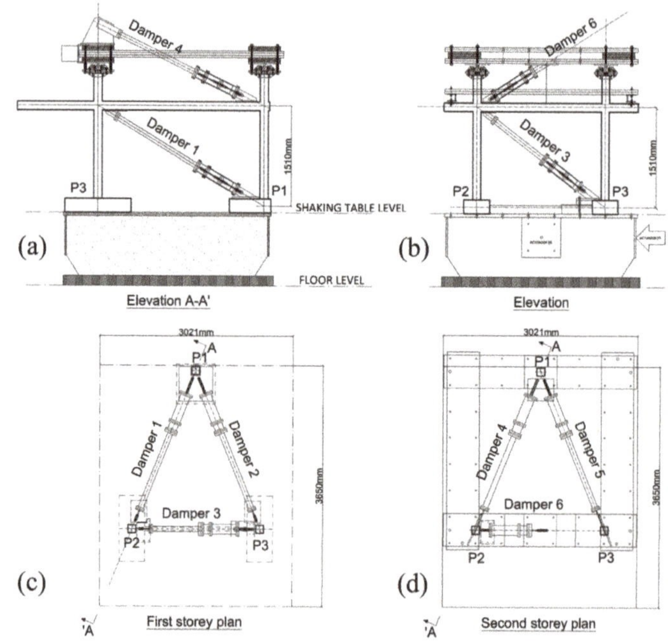

Figure 19. Details of the experimental set up for the shake table tests: (**a**) elevation view A-A′, (**b**) elevation, (**c**) plan (first storey), (**d**) plan (second storey).

4.3. Seismic Tests and Results

The test specimen was subjected simultaneously to the two horizontal components (NS and EW) of the far-field ground motion recorded at Calitri during the Campano Lucano (Italy, 1980) earthquake. The response on the NiTi dampers expressed in terms of axial force F and axial displacement δ was obtained from the measurements provided by the displacement transducers and the strain gauges. Figure 20 shows with dot lines the F-δ curves obtained for one of the NiTi bars. Meanwhile, a numerical model that represents the hysteretic behavior of the NiTi bar was developed, as explained in Section 3. The numerical model was subjected to the imposed axial displacements measured with the displacement transducers during the tests. The solid red line in Figure 20 shows the hysteresis curves predicted with the model put forth in Section 3, and the dot lines the experimental results obtained from the shake table tests. The proposed model is seen to predict with reasonable accuracy the forces sustained by the NiTi bar. In addition to the similarity in the shape of the hysteresis loops, the total amount of energy dissipated by the proposed model (E_D = 2699 kN·mm) is very similar to the actual value (E_D = 2563 kN·mm) measured during the tests. Finally, the response of the structure equipped with the SMA dampers is compared in Figure 21 with that of a counterpart structure without dampers that were subjected to the same earthquake in a previous study [28]. The response is compared in terms of maximum inter-story drift for the horizontal X direction, IDx, and in the Y direction IDy. The inter-story drift is defined as the relative lateral displacement between the top and bottom parts of each story, divided by the story height. It can be seen, that the SMA dampers reduced to less than one fourth the maximum inter-story drifts.

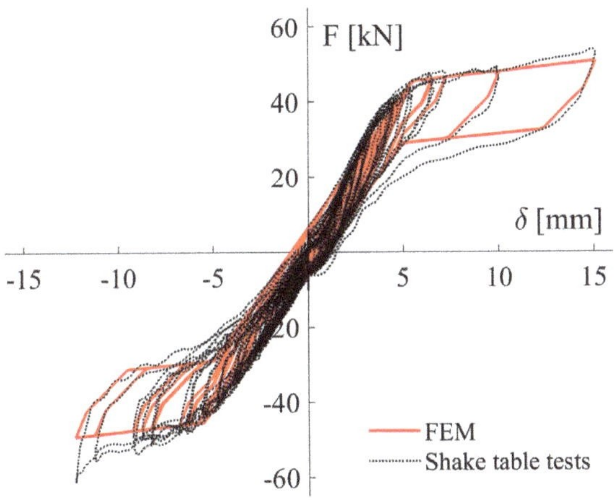

Figure 20. Experimental test vs numerical models implemented in FEM software.

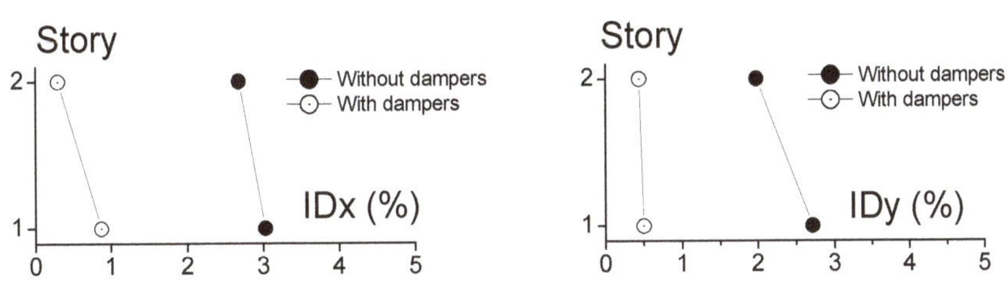

Figure 21. Response of the structure with and without SMA dampers.

5. Conclusions

A set of 12.7 mm diameter NiTi bars were tested under cyclic loading in both static and dynamic conditions, and at different frequencies. Some of them were tested up to failure in order to evaluate the ultimate energy dissipation capacity when subjected to low-cycle fatigue. Under cyclic static loadings at constant amplitude, the shape of the hysteresis loops stabilized after some cycles of deformation and the residual strain remained unaffected. The shape of the hysteretic loop is influenced by the frequency applied. Increasing the frequency leads to an increase in the loading and unloading stresses and a reduction of the amount of energy dissipated in each cycle, while the residual stress remains almost constant. The value of the equivalent viscous damping ratio was similar in all specimens tested, ranging between about 2.5% and 3% under dynamic loadings. The amount of energy dissipated under dynamic loading was 62% of the counterpart value obtained under static loads, and about 40% lower than the values reported in the literature for NiTi wires. The ultimate energy dissipation capacity of NiTi bars under low-cycle fatigue was found to be consistent with the values reported in the literature for NiTi bars subjected to high-cycle fatigue. An expression is proposed to quantify the ultimate energy dissipation capacity in terms of the number of cycles at constant amplitude to failure, and the normalized energy dissipation in one cycle. Finally, a new hysteretic model that consists of two springs connected in parallel is proposed for characterizing the hysteretic behavior of NiTi bars. One spring follows the conventional flag-type model (weighted by a factor γ) and the other spring follows an elastic-perfectly plastic rule. The proposed model is calibrated with the results of cyclic tests and validated with the results of dynamic shake table tests conducted on an RC structure equipped with NiTi bars as energy dissipation devices.

Author Contributions: Conceptualization, methodology, review, and editing, A.B.-C.; formal analysis, investigation and original draft preparation, G.G.-S.; test and validation G.G.-S., D.G.-L., and D.E.-M.

Funding: This work is funded by the European Union under the program H2020 with the project SERA "Seismology and Earthquake Engineering Research Infrastructure Alliance for Europe", responding to the priorities identified in the call INFRAIA-01-2016-2017 Research Infrastructure for Earthquake Hazard H2020-INFRAIA-2016-1.

Acknowledgments: One of the authors was given a grant from the Spanish Government FPU16/03006 that is gratefully acknowledged. Thanks are given to Jaime Gálvez Ruiz for helping with the management of the dynamic cyclic tests.

Conflicts of Interest: The authors declare no conflict of interest.

References

1. Dolce, M.; Cardone, D. Mechanical behaviour of shape memory alloys for seismic applications 1. Martensite and austenite NiTi bars subjected to torsion. *Int. J. Mech. Sci.* **2001**, *43*, 2631–2656. [CrossRef]
2. Dolce, M.; Cardone, D. Mechanical behaviour of shape memory alloys for seismic applications 2. Austenite NiTi wires subjected to tension. *Int. J. Mech. Sci.* **2001**, *43*, 2657–2677. [CrossRef]
3. DesRoches, R.; McCormick, J.; Delemont, M. Cyclic properties of superelastic shape memory alloy wires and bars. *J. Struct. Eng.-ASCE* **2004**, *130*, 38–46. [CrossRef]
4. McCormick, J.; DesRoches, R.; Fugazza, D.; Auricchio, F. Seismic vibration control using superelastic shape memory alloys. *J. Eng. Mater. Technol.-Trans. ASME.* **2006**, *128*, 294–301. [CrossRef]
5. McCormick, J.; Tyber, J.; DesRoches, R.; Gall, K.; Maier, H.J. Structural engineering with NiTi. II: Mechanical behavior and scaling. *J. Eng. Mech.-ASCE* **2007**, *133*, 1019–1029. [CrossRef]
6. Tyber, J.; McCormick, J.; Gall, K.; DesRoches, R.; Maier, H.J.; Maksoud, A.E.A. Structural engineering with NiTi. 1: Basic materials characterization. *J. Eng. Mech.-ASCE* **2007**, *133*, 1009–1018. [CrossRef]
7. Ozbulut, O.E.; Hurlebaus, S.; Desroches, R. Seismic response control using shape memory alloys: A review. *J. Intell. Mater. Syst. Struct.* **2011**, *22*, 1531–1549. [CrossRef]
8. Wang, J.; Zhao, H. High performance damage-resistant seismic resistant structural systems for sustainable and resilient city: A review. *Shock Vibrat.* **2018**, *2018*, 8703697. [CrossRef]
9. Lagoudas, D.C. *Shape Memory Alloys: Modeling and Engineering Applications*; Springer: New York, NY, USA, 2008; pp. 6–7.

10. Hartl, D.J.; Lagoudas, D.C. Aerospace applications of shape memory allows. *J. Aerosp. Eng.* **2007**, *221*, 535–552. [CrossRef]
11. Chopra, A.K. *Dynamics of Structures*; Prentice Hall: Upper Saddle River, NJ, USA, 1995; Volume 3.
12. Mahtabi, M.J.; Shamsaei, N.; Mitchell, M.R. Fatigue of nitinol: The state-of-the-art and ongoing challenges. *J. Mech. Behav. Biomed. Mater.* **2015**, *50*, 228–254. [CrossRef]
13. Song, D.; Kang, G.; Kang, Q.; Yu, C.; Zhang, C. Experimental observations on uniaxial whole-life transformation ratchetting and low-cycle stress fatigue of super-elastic NiTi shape memory alloy micro-tubes. *Smart Mater. Struct.* **2015**, *24*, 075004. [CrossRef]
14. Zhang, Y.; You, Y.; Moumni, Z.; Anlas, G.; Zhu, J.; Zhang, W. Experimental and theoretical investigation of the frequency effect on low cycle fatigue of shape memory alloys. *Int. J. Plast.* **2017**, *90*, 1–30. [CrossRef]
15. Eggeler, G.; Hornbogen, E.; Yawny, A.; Heckmann, A.; Wagner, M. Structural and functional fatigue of NiTi shape memory alloys. *Mater. Sci. Eng. A-Struct. Mater. Prop. Microstruct. Process.* **2004**, *378*, 24–33. [CrossRef]
16. Kimiecik, M.; Jones, J.W.; Daly, S. The effect of microstructure on stress-induced martensitic transformation under cyclic loading in the SMA nickel-titanium. *J. Mech. Phys. Solids* **2016**, *89*, 16–30. [CrossRef]
17. Zhu, S.; Zhang, Y. Loading rate effect on superelastic SMA-based seismic response modification devices. *Earthq. Struct.* **2013**, *4*, 607–627. [CrossRef]
18. Kan, Q.; Yu, C.; Kang, G.; Li, J.; Yan, W. Experimental observations on rate-dependent cyclic deformation of super-elastic NiTi shape memory alloy. *Mech. Mater.* **2016**, *97*, 48–58. [CrossRef]
19. Ammar, O.; Haddar, N.; Dieng, L. Experimental investigation of the pseudoelastic behaviour of NiTi wires under strain- and stress-controlled cyclic tensile loadings. *Intermetallics* **2017**, *81*, 52–61. [CrossRef]
20. Nemat-Nasser, S.; Choi, J.-Y.; Guo, W.-G.; Isaacs, J.B. Very high strain-rate response of a NiTi shape-memory alloy. *Mech. Mater.* **2005**, *37*, 287–298. [CrossRef]
21. Treadway, J.; Abolmaali, A.; Lu, F.; Aswath, P. Tensile and fatigue behavior of superelastic shape memory rods. *Mater. Des.* **2015**, *86*, 105–113. [CrossRef]
22. Kang, G.; Song, D. Review on structural fatigue of NiTi shape memory alloys: Pure mechanical and thermo-mechanical ones. *Theor. Appl. Mech. Lett.* **2015**, *5*, 245–254. [CrossRef]
23. Zhang, Y.; Zhu, J.; Moumni, Z.; Van Herpen, A.; Zhang, W. Energy-based fatigue model for shape memory alloys including thermomechanical coupling. *Smart Mater. Struct.* **2016**, *25*, 035042. [CrossRef]
24. Moumni, Z.; Van Herpen, A.; Riberty, P. Fatigue analysis of shape memory alloys: Energy approach. *Smart Mater. Struct.* **2005**, *14*, S287–S292. [CrossRef]
25. Ikeda, T.; Nae, F.A.; Naito, H.; Matsuzaki, Y. Constitutive model of shape memory alloys for unidirectional loading considering inner hysteresis loops. *Smart Mater. Struct.* **2004**, *13*, 916–925. [CrossRef]
26. Saleeb, A.F.; Padula II, S.A.; Kumar, A. A multi-axial, multimechanism based constitutive model for the comprehensive representation of the evolutionary response of SMAs under general thermomechanical loading conditions. *Int. J. Plast.* **2011**, *27*, 655–687. [CrossRef]
27. Karakalas, A.A.; Machairas, T.T.; Solomou, A.G.; Saravanos, D.A. Modeling of partial transformation cycles of SMAs with a modified hardening function. *Smart Mater. Struct.* **2019**, *28*, 1–20. [CrossRef]
28. Benavent-Climent, A.; Galé-Lamuela, D.; Donaire-Avila, J. Energy capacity and seismic performance of RC waffle-flat plate structures under two components of far-field ground motions: Shake table tests. *Earthquake Eng. Struct. Dyn.* **2019**, *48*, 949–969. [CrossRef]

MDPI
St. Alban-Anlage 66
4052 Basel
Switzerland
Tel. +41 61 683 77 34
Fax +41 61 302 89 18
www.mdpi.com

Metals Editorial Office
E-mail: metals@mdpi.com
www.mdpi.com/journal/metals

www.ingramcontent.com/pod-product-compliance
Lightning Source LLC
Chambersburg PA
CBHW062044220926
51850CB00010B/2658
* 9 7 8 3 0 3 6 5 2 4 7 0 2 *